AGRICULTURE ISSUES AND POLICIES

DISTILLERS' DRIED GRAINS

MARKET AND USAGE TRENDS

AGRICULTURE ISSUES AND POLICIES

Additional books in this series can be found on Nova's website under the Series tab.

Additional E-books in this series can be found on Nova's website under the E-book tab.

AGRICULTURE ISSUES AND POLICIES

DISTILLERS' DRIED GRAINS

MARKET AND USAGE TRENDS

JUSTIN N. SCHUBER
EDITOR

nova publishers
New York

For permission to use material from this book please contact us:
Telephone 631-231-7269; Fax 631-231-8175
Web Site: http://www.novapublishers.com

Additional color graphics may be available in the e-book version of this book.

Library of Congress Cataloging-in-Publication Data

ISBN: 978-1-62417-683-8

Published by Nova Science Publishers, Inc. † New York

CONTENTS

PREFACE

Growth in corn dry-mill ethanol production has surged in the past several years, simultaneously creating a co-product--distillers' grains (DDGS). Many in the U.S. feed industry were concerned about the size of this new feed source and whether it could be used entirely by the feed industry, but they also worried about the price discovery process for the product. This book provides an overview of the methodology used to estimate the U.S. supply and consumption of DDGS. Potential domestic and export use of U.S. DDGS exceeds current production and is likely to exceed future production as ethanol production continues to grow. This book discusses the DDGS price discovery process along with the price relationships of distillers' grains, corn, and soybean meal.

Chapter 1 - Growth in corn dry-mill ethanol production has surged in the past several years, simultaneously creating a coproduct—distillers' grains (DDGS). Many in the U.S. feed industry were concerned about the size of this new feed source and whether it could be used entirely by the feed industry, but they also worried about the price discovery process for the product. The authors of this report provide a transparent methodology to estimate U.S. supply and consumption of DDGS. Potential domestic and export use of U.S. DDGS exceeds current production and is likely to exceed future production as ethanol production continues to grow. The authors identify the DDGS price discovery process along with the price relationships of distillers' grains, corn, and soybean meal.

Chapter 2 - Corn-based dry-mill ethanol production and that of its coproducts—notably distillers' dried grains with solubles (DDGS)—has surged in the past several years. The U.S. feed industry has focused on the size of this new feed source and its impact on the U.S. feed market, particularly the degree that DDGS substitute for corn and soybean meal in livestock/poultry diets and reduce ethanol's impact on the feed market. This study develops a method to estimate the potential use of U.S. DDGS and its substitutability for corn and soybean meal in U.S. feed rations. Findings demonstrate that, in aggregate (including major types of livestock/poultry), a metric ton of DDGS can replace, on average, 1.22 metric tons of feed consisting of corn and soybean meal in the United States. Over time, DDGS may substitute for less corn and more soybean meal as the share of beef cattle consumption of DDGS declines slightly (although increasing in absolute terms), with offsetting share increases in dairy cattle, swine, and poultry. Feed market impacts of increased corn use for ethanol are smaller than that indicated by the total amount of corn used for ethanol production because of DDGS.

Chapter 3 - The expansion of corn-based ethanol production in the United States yields a large volume of residual co-products called distillers dried grain with solubles (DDGS). Most DDGS are utilized in the domestic U. S. market, but Chinese importers seeking raw materials for animal feed have emerged as a significant export market. High feed prices and favorable tax treatment within China stimulated a surge of imports of U. S. DDGS during 2009-11. China's potential demand for U. S. DDGS depends on various factors that include the price of corn, Chinese policy, and the availability and price of other substitute feed ingredients, such as the byproducts of grain processing in China (residual products from alcohol production). Demand is robust, but slower growth in the U. S. supply of DDGS and uncertainties about Chinese policy may constrain growth in exports to China.

In: Distillers' Dried Grains
Editor: Justin N. Schuber

ISBN: 978-1-62417-683-8

Chapter 1

MARKET ISSUES AND PROSPECTS FOR U.S. DISTILLERS' GRAINS SUPPLY, USE, AND PRICE RELATIONSHIPS*

Linwood Hoffman and Allen Baker

ABSTRACT

Growth in corn dry-mill ethanol production has surged in the past several years, simultaneously creating a coproduct—distillers' grains (DDGS).

Many in the U.S. feed industry were concerned about the size of this new feed source and whether it could be used entirely by the feed industry, but they also worried about the price discovery process for the product.

The authors of this report provide a transparent methodology to estimate U.S. supply and consumption of DDGS. Potential domestic and export use of U.S. DDGS exceeds current production and is likely to exceed future production as ethanol production continues to grow. The authors identify the DDGS price discovery process along with the price relationships of distillers' grains, corn, and soybean meal.

---

* This is an edited, reformatted and augmented version of United States Department of Agriculture, A Report from the Economic Research Service, Publication No. FDS-10k-01, dated December 2010.

INTRODUCTION

Growth in U.S. ethanol production seems certain for the near- to mid-term, stimulated, in part, by longer run energy prices and the 2005 and 2007 Energy Acts.[1]

In 2005/06, 4.5 billion gallons of ethanol were produced. By 2008/09, production more than doubled to 10.2 billion gallons, with a corresponding increase in distillers' grains (DDGS),[2] a coproduct of ethanol production that can be used as a livestock feed.[3]

Prices for corn, a feedstock for both ethanol and livestock feed, and other feed-related products have risen, causing concern for the financial viability of livestock and poultry producers.

The use of distillers' grains continues to expand and partially replace corn and soybean meal in the U.S. feed market. DDGS can supply both energy and protein in livestock and poultry rations, but use may be limited due to nutritional or price considerations. Prices of DDGS relative to alternative feed ingredients would need to adjust to reflect its nutrient content.

As the industry expands, producers may wonder about the price relationships between DDGS, corn, and soybean meal.

Distillers' grains have long been substituted for corn and soybean meal, but as the market expands, questions remain over the substitution rate of DDGS for energy (corn) and protein (soybean meal) in livestock and poultry rations. As the market expands, industry analysts may be concerned about:

- The relative balance between supply and disappearance, regardless of increased DDGS feeding familiarity by feeders and the improvement of DDGS product quality (Cooper, 2006);
- The supply of U.S. DDGS exceeding feed-use potential;
- The potential use of DDGS for domestic or export purposes; and
- Current or future U.S. production of DDGS exceeding potential uses.

The growth in this new feed market reinforces the need for a transparent methodology to estimate supply and consumption for the U.S. distillers' grain market. Currently, production estimates are made by the U.S. Census Bureau (U.S. Department of Commerce (a)), the Renewable Fuels Association (*Ethanol Industry Outlook* for 2008, 2009, and 2010), and other research organizations.

The U.S. Census Bureau began providing estimates of distillers' grain production in early 2007. These production estimates were considered incomplete by many Government and industry analysts. Other production estimates lack transparency and comparability. So, until the U.S. Census Bureau estimates are more complete and reflect ethanol production data, a transparent estimation method would serve both the Government and the industry, offering estimates for the production and consumption of distillers' grains.

Data on ethanol coproduct feed use (distillers' grains, corn gluten feed, and corn gluten meal) will be valuable for those conducting analytical analyses of the U.S. feed sector (e.g., Ferris, 2006).

This report details the development of the U.S. distillers' grains market and provides a transparent methodology (see "Appendix: Estimation of U.S. Distillers' Grains Supply and Use") to estimate supply and use of distillers' grains.

The authors of this study examined price discovery and risk management of distillers' grains along with the price relationships of DDGS, corn, and soybean meal.

ESTIMATING U.S. DISTILLERS' GRAIN SUPPLY AND USE

The surge in ethanol production over the last several years has been paralleled by a surge in the production of distillers' grains. Currently, consistent data do not exist to measure this rapidly growing segment of the feed industry. We offer a methodology that produces a transparent source of information to gauge the supply and use of distillers' grains (table 1).

For production, we estimated the amount of corn used by beverage distilleries and the resulting amount of distillers' grains produced from this process. We also estimated the amount of corn used by dry-mill ethanol plants and the resulting distillers' grains.

We then added imports of distillers' grains based on trade data from the U.S. Department of Commerce (b) and from the Economic Research Service's Feed Grains Database (2010).

By combining these components, we estimated the total supply of distillers' grains. From this supply, we subtracted exports based on trade data from the same two sources to arrive at feed and residual use. Details of this estimation procedure can be found in the appendix.

Table 1. Supply and disappearance of corn-based distillers' grains (DDGS)

	Column 1		Column 2				Column 3			Column 4		Column 5		
	Supply									Disappearance				
	Production						Imports		Total	Feed and residual		Exports		Total
Marketing year	Distillers' spent grains from beverage distilleries		Distillers' spent grains from dry mill plants producing ethanol[1]		Total		Brewers' and distillers' dregs and wastes[2,3]					Brewers' and distillers' dregs and wastes[2,3]		
	Million metric tons	*Percent of total production*	*Million metric tons*	*Percent of total production*	*Million metric tons*	*Percent of total supply*	*Million metric tons*	*Percent of total supply*	*Million metric tons*	*Million metric tons*	*Percent of total disappearance*	*Million metric tons*	*Percent of total disappearance*	*Million metric tons*
1992/93	0.8	40	1.2	60	2.0	97	0.1	3	2.1	2.1	100	0.0	0	2.1
1993/94	0.6	27	1.6	73	2.2	96	0.1	4	2.3	2.1	92	0.2	8	2.3
1994/95	0.5	35	1.0	65	1.5	94	0.1	6	1.6	0.9	55	0.7	45	1.6
1995/96	0.8	77	0.2	23	1.0	93	0.1	7	1.1	0.5	41	0.6	59	1.1
1996/97	0.8	64	0.5	36	1.3	91	0.1	9	1.4	0.8	53	0.7	47	1.4
1997/98	0.9	52	0.8	48	1.6	94	0.1	6	1.7	1.1	65	0.6	35	1.7
1998/99	0.8	47	1.0	53	1.8	93	0.1	7	1.9	1.2	62	0.7	38	1.9
1999/00	0.9	51	0.9	49	1.7	92	0.1	8	1.9	1.1	59	0.8	41	1.9
2000/01	0.9	36	1.6	64	2.5	95	0.1	5	2.6	1.8	70	0.8	30	2.6
2001/02	0.9	31	2.0	69	2.9	96	0.1	4	3.0	2.1	72	0.9	28	3.0
2002/03	0.9	17	4.3	83	5.2	98	0.1	2	5.3	4.6	86	0.8	14	5.3
2003/04	0.9	13	6.1	87	7.0	99	0.1	1	7.1	6.4	90	0.7	10	7.1
2004/05	0.9	11	7.3	89	8.2	99	0.1	1	8.3	7.4	88	1.0	12	8.3
2005/06	0.9	9	9.5	91	10.4	99	0.1	1	10.5	9.3	88	1.2	12	10.5
2006/07	0.9	7	13.2	93	14.1	99	0.2	1	14.3	12.5	88	1.8	12	14.3
2007/08	0.9	4	20.5	96	21.4	100	0.1	0	21.5	17.6	82	3.9	18	21.5
2008/09	0.9	3	25.6	97	26.5	99	0.3	1	26.8	21.8	81	5.0	19	26.8
2009/10[4]	0.9	3	32.0	97	32.9	99	0.4	1	33.3	25.0	75	8.3	25	33.3
2010/11[4]	0.9	3	33.2	97	34.1	99	0.5	1	34.6	25.9	75	8.7	25	34.6

[1] Does not account for noncorn spent grains.
[2] Assumes brewers' spent grains are minor.
[3] May contain noncorn brewers' and distillers' dregs and wastes.
[4] Estimates are from the September 10, 2010 World Agricultural Supply and Demand Estimates and Feed Grains Database.

Source: Distillers' spent grains from beverage distilleries are from U.S. Department of the Treasury, Alcohol and Tobacco Tax and Trade Bureau; monthly statistical reports, 1992-2010, http://www.ttb.gov/statistics/alcohol_stats.shtml#alcoholfuel; distillers' spent grains from dry-mill ethanol plants are from table 3 of this report (corn used for dry-mill ethanol plants multiplied times 17.5 pounds of distillers' grains per bushel of corn processed for 1992/93 through 2006/07 and 17.4 pounds of distillers' grains per bushel of corn processed for 2007/08 through 2010/11); imports and exports are taken from USDA, Economic Research Service, Feed Grains Database or U.S. Department of Commerce (b).

TOTAL SUPPLY OF U.S. DISTILLERS' GRAINS ACCELERATES OVER THE PAST 4 YEARS

The estimated supply of U.S. distillers' grains is expected to total 33.3 million metric tons for 2009/10, more than four times that in 2003/04 (see table 1) (see also, "Appendix: Estimation of U.S. Distillers' Grains Supply and Use" for an explanation of the methodology that provided these estimates.) Most distillers' grains come from dry-mill fuel ethanol production and a lesser amount comes from dry-mill beverage distilleries. Imports are a minor segment of the market.[4] Most distillers' grains are derived from corn with smaller amounts coming from sorghum and wheat. Changes in dry-mill ethanol technology may further alter the composition of distillers' grains. Growth in distillers' grain production will continue to follow corn use for ethanol production, but both are expected to slow compared with the rapid expansion over the past several years (USDA, February 2010).

Long History of Using DDGS as a Livestock Feed

Distillers' grains, a byproduct of dry-mill beverage and ethanol fuel production, have long been fed to livestock. In the late 19th century, the alcoholic beverage distilling industry pioneered the act of recovering nutrients from grains that had undergone fermentation (Schingoethe, 2004). Distillers' grains are the spent coproducts of the fermentation and distillation process. These distillers' grains were recognized as an excellent feed for dairy cattle.[5]

Ethyl alcohol was critical for the manufacture of munitions during World War II, and the beverage industry (dry-mills) met most of this demand (Weiss et al., 2007). Since then, several factors have contributed to the expansion of U.S. ethanol fuel production from both wet- and dry-mill processing plants:

- The world oil crisis in the 1970s;
- State and Federal subsidies for ethanol production;
- Clean air legislation in the early 1990s;
- Ethanol's use as an octane enhancer;
- The replacement of methyl tertiary butyl ether (MTBE); and
- Rising energy costs and mandates for biofuel use in the Energy Acts of 2005 and 2007.

Corn Is Main Feedstock for Ethanol Production

Corn is the primary grain used in wet- and dry-mill ethanol plants (accounting for about 98 percent of all ethanol feedstocks) because its starch content is highly fermentable. Some ethanol plants use sorghum or a blend of corn with barley, wheat, or sorghum. Other plants may use excess whey, beverage, or sugar wastes as a feedstock. The decision to use a particular feedstock depends upon the geographical location, costs, and availability of other feedstocks relative to corn. In the future, biomass from other sources (e.g., grasses, wood pulp, or crop residue) may contribute more feedstock for ethanol production (Coyle, 2010).

Different Feed Coproducts from Wet Versus Dry Processes

Grain-based fuel ethanol production results from two main processes—wet and dry corn milling. Although ethanol production changes corn's feed value by removing most of the starch content, the remaining ethanol coproduct retains the following nutrients: protein, minerals, fat, fiber, and vitamins. Because ethanol production processes differ, the coproducts also differ. For example, wet corn mills produce corn gluten feed and corn gluten meal. Dry corn mills produce distillers' spent grains. More recently, fractionation technologies used by some plants has begun to change the composition of distillers' spent grains from the dry-mill process.[6]

Wet Milling

The wet-milling process removes the maximum amount of starch from the kernel by first adding water to the grain and allowing it to steep so the starch can be removed, along with the corn bran, corn gluten meal (protein), germ, and soluble components. Since this process separates the germ (from which corn oil is extracted), corn oil is one coproduct produced from the wet-milling process regardless of other products produced, including starch, high fructose corn syrup (HFCS), glucose/ dextrose, and/or ethanol. Feed byproducts from the wet process include corn gluten feed, corn gluten meal, and corn germ meal.7,[8] These coproducts can be fed to livestock in either wet or dry form. Corn gluten feed usually consists of corn bran, steeping liquid, and germ meal. One bushel of corn processed by a wet mill yields about 12.9 pounds of corn gluten feed and 3.1 pounds of corn gluten meal (Watson, 1977), plus carbon dioxide, and about 2.7 gallons of ethanol (Rendleman and Shapouri, 2007).

The protein content of corn gluten feed is usually about 21 percent (as fed), while corn gluten meal's protein content is around 60 percent (as fed).[9] We estimate that 475.1 million bushels of corn were used by corn wet mills to produce fuel ethanol during 2009/10 (table 2) (see also, "Appendix: Estimation of U.S. Distillers' Grains Supply and Use" for an explanation of the methodology that provided these estimates).

Dry Milling

The dry milling process (mash distillation) cleans and grinds the grain to reduce the particle size; usually the entire corn kernel is used in fermentation. Corn is about two-thirds starch, which is converted into ethanol and carbon dioxide during the distillation and fermentation process. The remaining third consists of distillers' grains. A bushel of corn processed into ethanol by dry mills produces approximately 17.5 pounds of distillers' spent grains, carbon dioxide, and 2.7 gallons of ethanol. The nutrient content (protein, fat, minerals, and vitamins) of the spent grains increases by three times because prior to fermentation these nutrients represented only about a third of corn's dry matter.[10]

While there are several different forms of distillers' coproducts from dry-mill ethanol plants, most of the distillers' spent grains are dried and sold as distillers' dried grains with solubles (DDGS). The liquid removed from the mash is called thin stillage. Some of the thin stillage can be directed back to the cooking and distillation process to reduce the amount of fresh water required by the cooking and distillation process, and the remainder can be sold directly as high-moisture cattle feed or dehydrated to produce condensed distiller's solubles (CDS).

The residual solids (coarse grains fraction) are referred to as wet distillers' grains and can be used as cattle feed or dried to produce dried distillers' grains (DDG). CDS can be used as cattle feed or blended with distillers' grains to produce DDGS. Distillers' grains with solubles are sold wet (WDGS; 30 to 35 percent dry matter), modified (MDGS; 45 to 50 percent dry matter), or dried (DDGS; 88 to 90 percent dry matter).

Distillers' grains can be fed either wet or dry. Approximately 36 percent of the distillers' grains with solubles fed in the United States are wet and 64 percent are dry (Renewable Fuels Association, *Ethanol Industry Outlook*, 2009). The authors of this report will focus specifically on DDGS and convert all distillers' grains into an equivalent DDGS weight.

Table 2. Estimation of U.S. corn use for wet-mill ethanol production

Crop year	Column 1	Column 2	Column 3	Column 4	Column 5	Column 6	Column 7	Column 8
	Corn oil production		Corn use	Corn used to produce various wet-mill products				
	Total[1]	Adjusted for dry mill products[2]	for wet-mill corn oil production[3]	High fructose corn syrup	Glucose/ dextrose	Starch	Sum of three products	Ethanol
	Thousand pounds		Million bushels					
1992/93	1,866,203	1,834,478	1,118.6	414.8	214.3	217.6	846.7	271.9
1993/94	1,906,233	1,873,827	1,142.6	441.0	219.2	224.9	885.1	257.5
1994/95	2,201,751	2,164,321	1,319.7	458.6	224.5	230.4	913.6	406.1
1995/96	2,155,459	2,118,816	1,292.0	472.9	227.3	226.1	926.3	365.7
1996/97	2,223,434	2,185,636	1,332.7	492.3	232.9	238.3	963.4	369.3
1997/98	2,328,157	2,288,578	1,395.5	528.0	236.5	248.1	1,012.6	382.9
1998/99	2,366,952	2,326,714	1,418.7	544.0	226.3	242.5	1,012.8	405.9
1999/00	2,491,224	2,448,873	1,493.2	551.9	229.3	253.4	1,034.6	458.6
2000/01	2,406,507	2,365,596	1,442.4	536.2	226.7	250.0	1,012.9	429.5
2001/02	2,453,609	2,411,898	1,470.7	542.2	226.5	248.8	1,017.5	453.1
2002/03	2,460,883	2,419,048	1,475.0	532.2	231.0	258.2	1,021.4	453.6
2003/04	2,399,561	2,358,768	1,438.3	529.8	238.3	273.3	1,041.4	396.9
2004/05	2,407,167	2,366,245	1,442.8	525.1	234.5	282.0	1,041.6	401.2
2005/06	2,462,773	2,420,906	1,476.2	545.0	245.1	280.4	1,070.5	405.7
2006/07	2,559,870	2,511,232	1,531.2	535.3	259.4	276.9	1,071.6	459.6
2007/08	2,508,917	2,458,739	1,499.2	522.8	256.0	264.7	1,043.5	455.7
2008/09	2,409,981	2,355,756	1,436.4	489.1	245.0	234.1	968.2	468.2
2009/10[4]	2,500,000	2,443,750	1,490.1	515.0	255.0	245.0	1,015.0	475.1
2010/11[4]	2,575,000	2,517,063	1,534.8	525.0	260.0	250.0	1,035.0	499.8

[1] Historical corn oil production data from the U.S. Department of Commerce (a), Census Bureau, Fats and Oils: Oilseed Crushings Circular, M311J, table 2, and from Corn Oil, http://www.census.gov/manufacturing/cir/historical_ data/ m311j/index.html Adjusted to corn crop year, September-August.

[2] Corn oil projections for 1992/93-2005/06 are reduced by 1.7 percent to account for dry-mill corn oil production. Reductions for 06/07 = 1.9 percent; 07/08 = 2.0 percent; 08/09 = 2.25 percent; 09/10 = 2.25 percent; and 10/11 = 2.25 percent to account for fractionation oil removal.

[3] Assumes corn oil yield of 1.64 pounds/bushel.

[4] Corn oil projections from the September 13, 2010, Oil Crops Outlook; wet-mill products, excluding ethanol, from the September 14, 2010, Feed Grains Outlook, p. 15.

Source: USDA, Economic Research Service calculations based on data from U.S. Department of Commerce (a), Census Bureau, Fats and Oils: Oilseed Crushings Circular; Economic Research Service Oil Crops Outlook.

Supplies of Ethanol Coproducts Rapidly Expanding

Most of the expansion in ethanol fuel production has come from dry-mill plants with little additional capacity expansion from wet mills. Wet mills are more expensive to build, more versatile in the products they can produce, yield slightly less ethanol per bushel, and offer more valuable coproducts. Wet mills used to account for most of the U.S. ethanol production, but new construction has shifted to dry mills, partly because dry mills are less costly to build. For 2009/10, we estimate that 11 percent of U.S. fuel ethanol was produced in wet mills with the remaining 89 percent produced in dry mills, compared with 2001/02 when wet mills produced 64 percent and dry mills 36 percent of total ethanol production (table 3).

Table 3. Estimation of corn use for dry-mill ethanol production

Crop year	Column 1 Total corn use for fuel ethanol production[1]	Column 2 Corn use for wet-mill ethanol production[2]	Column 3 Corn use for dry-mill ethanol production[2]	Percent used by mill type: Wet mills	Dry mills	Total
		Million bushels			*Percent*	
1992/93	425.5	271.9	153.6	63.9	36.1	100
1993/94	458.3	257.5	200.7	56.2	43.8	100
1994/95	532.8	406.1	126.6	76.2	23.8	100
1995/96	395.7	365.7	30.0	92.4	7.6	100
1996/97	428.7	369.3	59.4	86.1	13.9	100
1997/98	481.1	382.9	98.2	79.6	20.4	100
1998/99	525.8	405.9	119.9	77.2	22.8	100
1999/00	565.8	458.6	107.1	81.1	18.9	100
2000/01	627.6	429.5	198.1	68.4	31.6	100
2001/02	705.9	453.1	252.8	64.2	35.8	100
2002/03	995.5	453.6	541.9	45.6	54.4	100
2003/04	1,167.5	396.9	770.7	34.0	66.0	100
2004/05	1,323.1	401.2	921.8	30.3	69.7	100
2005/06	1,602.8	405.7	1,197.1	25.3	74.7	100
2006/07	2,119.0	459.6	1,659.4	21.7	78.3	100
2007/08	3,049.0	455.7	2,593.3	14.9	85.1	100
2008/09	3,709.0	468.2	3,240.8	12.6	87.4	100
2009/10[3]	4,535.0	475.1	4,059.9	10.5	89.5	100
2010/11[3]	4,700.0	499.8	4,200.2	10.6	89.4	100

[1] Based on USDA, Economic Research Service, Feed Grains Database.

[2] Based on table 2 of this report. [3]Corn use for ethanol production from September 10, 2010, *World Agricultural Supply and Demand Estimates.*

Source: USDA, Economic Research Service, Feed Grains Database; table 2 of this report.

Production of distillers' grains, a feed coproduct from corn dry-mill ethanol production, surpassed feed coproducts from the wet-milling process (namely corn gluten feed and corn gluten meal) in marketing year 2006/07 (table 4).[11] Between 1992/93 and 2001/02, exports accounted for most of corn gluten feed's disappearance, but export's recent share of corn gluten feed has declined. This decline in corn gluten feed was due to the European Union's Common Agricultural Policy (CAP) reform and from the adoption of genetically modified corn in the United States. For example, cereal prices declined when CAP reform stimulated consumption of domestic feed. EU reforms also reduced the EU cow herd, which reduced the need for feed. Furthermore, U.S. corn gluten feed and its derivation from genetically modified corn may have precluded its use in EU feed rations.[12] This biotechnology concern may be present regardless of whether U.S. corn wet mills used only EU-approved corn varieties, due to fear of contamination of unapproved varieties. Tables 5 and 6 provide estimates of U.S. corn gluten feed and meal supply and disappearance.

As the production of corn-based ethanol increases, DDGS output increases because of the fixed proportions of ethanol and coproduct yield from corn bushels processed. For example, when dry-mill fuel ethanol production began to accelerate between 2002/03 and 2009/10 and dry-mill corn use rose from an estimated 0.5 to 4.1 billion bushels (see table 3), production of DDGS rose from an estimated 4.3 to 32.0 mmt (million metric tons) (see table 1).

Composition of Distillers' Grains Is Beginning to Change

In addition to the switch from wet to dry milling, new technologies and production processes are changing the composition of distillers' grains. When the ethanol boom started in the mid-2000s, essentially all additional capacity expansion came from dry-mill ethanol plants. These plants required less capital to build, and investors at the time were interested primarily in ethanol since its margins were more attractive (Randleman and Shapouri, 2007). Consequently, fewer investors were interested in the potential coproduct revenue stream. With time, however, ethanol margins tightened and ethanol plant operators looked for additional revenue from their coproducts by improving the quality and consistency of DDGS. Potential coproducts include carbon dioxide and, more recently, corn oil.[13] As processors extract as much value from the corn as possible, future spent grains from ethanol production could be used as an alternative feedstock for cellulosic ethanol, fuel for the distillation process or electricity generation, or with anaerobic digestion, assuming profitability supports this feedstock selection.

Presently, there are two dry-mill fractionation processes that can alter the composition of distillers' grains. The first process can be and has been added to the front end of dry-mill ethanol plants.[14] This process can be a wet or dry process and can produce bran, germ (whole or defatted germ meal), and various combinations of these prefermentation byproducts. For example, adding germ to the bran produces "high fat hominy." Separation of the germ (fat-rich portion of the corn kernel) and the high fiber bran from the endosperm prior to fermentation results in stillage that is lower in fiber, fat, and phosphorus and higher in protein than traditional distillers' grains (Onetti and Schwab, 2008). The starch content of these prefermentation byproducts also varies significantly, depending on the effectiveness of the different milling processes in separating the endosperm from the other kernel components prior to fermentation.

The second fractionation process, commonly called back-end, is a less costly endeavor than the front-end process and is rapidly gaining more widespread usage in the ethanol industry. The back-end process extracts corn oil from either stillage or CDS. Generally, the oil content found in DDGS is about 11.2 percent and can be reduced by a third to 7.5 percent. A more expensive technology, however, can remove from two-thirds to three-quarters of the oil, resulting in 3.7 percent to 2.8 percent oil content.

Generally, many of these new corn coproducts are higher in crude protein and crude fiber than "regular" DDGS and lower in crude fat. Although the amino acid concentration may increase somewhat in many of these high-protein coproducts, the protein quality (amino acid balance) is still poor relative to the requirements of monogastric animals.[15] Additionally, the reduced fat and increased fiber content of these new coproducts may provide lower energy value for swine and poultry (Shurson and Alghandi, 2008). Thus, the feeding and economic value of these "reduced fat" DDGS may be lower for swine and poultry, compared with "regular" DDGS. The nutrient composition of these coproducts, however, would likely have greater value in ruminant diets because the amino acid balance of corn protein is not as critical for ruminants as it is for swine and poultry. The increased amount of readily digestible fiber can provide a good source of energy, and the lower fat content may allow higher dietary inclusion rates for lactating dairy cows, avoiding milk-fat depression concerns. Until further evidence is provided, animal nutritionists have cautioned feeders about these new coproducts since the number of published studies evaluating them in livestock and poultry diets has been limited (Shurson and Alghamdi, 2008).

Growth of U.S. Distillers' Grains Supply Expected to Moderate

The U.S. Department of Agriculture projects that 5.0 billion bushels of corn will be used in fuel ethanol production in 2019/20 (USDA, February 2010), up from the 4.5 billion bushels projected for 2009/10 (USDA, September 2010). Slower annual growth for corn-based ethanol is projected, following the rapid expansion seen in recent years.[16] Consequently, the supply of distillers' grains is estimated to rise from about 33.3 mmt in 2009/10 to about 38.6 mmt in 2019/20.17,[18] Whether this projected production estimate represents a saturation of the feed market will be discussed later.

Consumption of Distillers' Grains Increases, Export Share Expands

The estimated disappearance of U.S. distillers' grains is expected to total 33.3 mmt for 2009/10, more than doubling since 2006/07 (see table 1). U.S. feed and residual use accounts for the largest share of consumption of U.S. distillers' grains. Domestic feeders have been rapidly adopting this product to take advantage of its nutrient and cost advantages. Despite the many issues of product quality, consistency, handling, and feed safety, DDGS provide good value per unit of energy and protein. Despite increased feeding of DDGS by domestic feeders, the rate of growth for exports exceeds the rate of growth in domestic use. Between marketing years 2003/04 and 2009/10, the domestic use share has dropped by 15 percentage points, with an offsetting increase in export share of U.S. distillers' grains disappearance.

Export share is increasing as foreign feeders adopt DDGS as a cheaper alternative to corn and U.S. trade groups work to alleviate concerns over product quality, consistency, and safety (Voegele, 2009).19, [20] While the European Union had been the largest importer of U.S. distillers' grains through 2005/06, CAP reform and biotechnology regulations have significantly reduced the quantity of U.S. exports to the EU market.[21] Since then, Mexico has become the largest importer of U.S. distillers' grains, followed by Canada and Asian countries (table 7) (Fox, 2008 and 2009; Paulson, 2008). The potential for increased exports to China appears promising (Fabiosa et al., 2009). For the 2009/10 marketing year, DDGS exports to China have been expanding and China could become the largest importer of U.S. distillers' grains.

Table 4. Estimates of feed coproducts from corn wet- and dry-mill related production processes

	Corn wet-mill production							Corn dry-mill production		
	HFCS, Glucose/dextrose, and starch		Ethanol		Total			Beverage ethanol	Fuel ethanol	Total
Crop year	Corn gluten:		Corn gluten:		Corn gluten:		Grand total	Distillers' grains produced from beverage and fuel production		
	Feed	Meal	Feed	Meal	Feed	Meal				
	Million metric tons									
1992/93	5.0	1.2	1.6	0.4	6.5	1.6	8.1	0.8	1.2	2.0
1993/94	5.2	1.2	1.5	0.4	6.7	1.6	8.3	0.6	1.6	2.2
1994/95	5.3	1.3	2.4	0.6	7.7	1.9	9.6	0.5	1.0	1.5
1995/96	5.4	1.3	2.1	0.5	7.6	1.8	9.4	0.8	0.2	1.0
1996/97	5.6	1.4	2.2	0.5	7.8	1.9	9.7	0.8	0.5	1.3
1997/98	5.9	1.4	2.2	0.5	8.2	2.0	10.1	0.9	0.8	1.6
1998/99	5.9	1.4	2.4	0.6	8.3	2.0	10.3	0.8	1.0	1.8
1999/00	6.1	1.5	2.7	0.6	8.7	2.1	10.8	0.9	0.9	1.7
2000/01	5.9	1.4	2.5	0.6	8.4	2.0	10.5	0.9	1.6	2.5
2001/02	6.0	1.4	2.7	0.6	8.6	2.1	10.7	0.9	2.0	2.9
2002/03	6.0	1.4	2.7	0.6	8.6	2.1	10.7	0.9	4.3	5.2
2003/04	6.1	1.5	2.3	0.6	8.4	2.0	10.4	0.9	6.1	7.0
2004/05	6.1	1.5	2.3	0.6	8.4	2.0	10.5	0.9	7.3	8.2
2005/06	6.3	1.5	2.4	0.6	8.6	2.1	10.7	0.9	9.5	10.4
2006/07	6.3	1.5	2.7	0.6	9.0	2.2	11.1	0.9	13.2	14.1
2007/08	6.1	1.5	2.7	0.6	8.8	2.1	10.9	0.9	20.5	21.4
2008/09	5.7	1.4	2.7	0.7	8.4	2.0	10.4	0.9	25.6	26.5
2009/10[1]	5.9	1.4	2.8	0.7	8.7	2.1	10.8	0.9	32.0	32.9
2010/11[1]	6.1	1.5	2.9	0.7	9.0	2.2	11.1	0.9	33.2	34.1

HFCS=High fructose corn syrup.

[1]Based on September 10, 2010, World Agricultural Supply and Demand Estimates.

Source: Data for corn dry-mill plants from table 1 of this report; data for corn wet-mill plants from table 2 of this report.

Table 5. Supply and disappearance of corn gluten feed

Marketing year	Supply					Disappearance				
	Production		Imports		Total	Feed and residual		Exports		Total
	Million metric tons	*Percent of total supply*	*Million metric tons*	*Percent of total supply*	*Million metric tons*	*Million metric tons*	*Percent of total disappearance*	*Million metric tons*	*Percent of total disappearance*	*Million metric tons*
1992/93	6.5	98.4	0.11	1.6	6.7	0.9	14.0	5.7	86.0	6.7
1993/94	6.7	98.6	0.10	1.4	6.8	1.1	16.6	5.7	83.4	6.8
1994/95	7.7	98.9	0.08	1.1	7.8	1.6	20.3	6.2	79.7	7.8
1995/96	7.6	98.8	0.10	1.2	7.7	2.2	28.3	5.5	71.7	7.7
1996/97	7.8	98.8	0.10	1.2	7.9	2.7	33.8	5.2	66.2	7.9
1997/98	8.2	99.1	0.07	0.9	8.2	3.3	39.9	4.9	60.1	8.2
1998/99	8.3	99.8	0.02	0.2	8.3	3.3	40.2	5.0	59.8	8.3
1999/00	8.7	99.8	0.02	0.2	8.8	3.6	41.2	5.1	58.8	8.8
2000/01	8.4	99.8	0.02	0.2	8.5	3.8	45.1	4.6	54.9	8.5
2001/02	8.6	99.8	0.02	0.2	8.6	4.1	47.0	4.6	53.0	8.6
2002/03	8.6	99.6	0.03	0.4	8.7	4.9	56.6	3.8	43.4	8.7
2003/04	8.4	99.5	0.05	0.5	8.5	4.9	58.4	3.5	41.6	8.5
2004/05	8.4	99.6	0.03	0.4	8.5	5.7	67.1	2.8	32.9	8.5
2005/06	8.6	99.6	0.03	0.4	8.7	6.0	68.7	2.7	31.3	8.7
2006/07	9.0	99.7	0.03	0.3	9.0	7.4	81.9	1.6	18.1	9.0
2007/08	8.8	99.4	0.05	0.6	8.8	7.4	83.6	1.4	16.4	8.8
2008/09	8.4	99.4	0.05	0.6	8.5	7.6	89.4	0.9	10.6	8.5
2009/10	8.7	99.4	0.05	0.6	8.8	7.9	89.7	0.9	10.3	8.8
2010/11	9.0	99.4	0.05	0.6	9.1	8.2	90.1	0.9	9.9	9.1

Source: Production numbers are from table 4 of this report; import and export data from U.S. Department of Commerce (b), Census Bureau, http://www.usatra deonline.gov/.

Table 6. Supply and disappearance of corn gluten meal

Marketing year	Supply					Disappearance				
	Production		Imports		Total	Feed and residual		Exports		Total
	Million metric tons	*Percent of total supply*	*Million metric tons*	*Percent of total supply*	*Million metric tons*	*Million metric tons*	*Percent of total disappearance*	*Million metric tons*	*Percent of total disappearance*	*Million metric tons*
1992/93	1.6	99.7	0.00	0.3	1.6	1.0	64.5	0.6	35.5	1.6
1993/94	1.6	99.6	0.01	0.4	1.6	0.9	54.4	0.7	45.6	1.6
1994/95	1.9	99.5	0.01	0.5	1.9	1.2	66.2	0.6	33.8	1.9
1995/96	1.8	99.2	0.01	0.8	1.8	1.1	57.9	0.8	42.1	1.8
1996/97	1.9	99.3	0.01	0.7	1.9	1.1	58.8	0.8	41.2	1.9
1997/98	2.0	99.4	0.01	0.6	2.0	1.2	61.1	0.8	38.9	2.0
1998/99	2.0	99.4	0.01	0.6	2.0	1.4	67.8	0.6	32.2	2.0
1999/00	2.1	99.4	0.01	0.6	2.1	1.4	67.1	0.7	32.9	2.1
2000/01	2.0	99.5	0.01	0.5	2.0	1.2	61.2	0.8	38.8	2.0
2001/02	2.1	99.3	0.01	0.7	2.1	1.3	60.7	0.8	39.3	2.1
2002/03	2.1	99.3	0.01	0.7	2.1	1.3	61.0	0.8	39.0	2.1
2003/04	2.0	99.4	0.01	0.6	2.0	1.1	55.6	0.9	44.4	2.0
2004/05	2.0	99.5	0.01	0.5	2.0	1.2	57.7	0.9	42.3	2.0
2005/06	2.1	99.4	0.01	0.6	2.1	1.2	56.8	0.9	43.2	2.1
2006/07	2.2	99.5	0.01	0.5	2.2	1.2	54.3	1.0	45.7	2.2
2007/08	2.1	99.4	0.01	0.6	2.1	1.0	48.2	1.1	51.8	2.1
2008/09	2.0	99.6	0.01	0.4	2.0	1.3	65.5	0.7	34.5	2.0
2009/10	2.1	99.5	0.01	0.5	2.1	1.4	66.8	0.7	33.2	2.1
2010/11	2.2	99.5	0.01	0.5	2.2	1.5	68.3	0.7	31.7	2.2

Source: Production numbers are from table 4 of this report; import and export data from U.S. Department of Commerce (b), Census Bureau, http://www.usatradeonline.gov/.

Table 7. Exports of U.S. brewers' or distillers' dregs and waste, by crop year

Area	Importing country	Crop year 1992/93	1993/94	1994/95	1995/96	1996/97	1997/98	1998/99	1999/00
		Metric tons							
North America	Canada	16,002	14,675	16,121	11,483	16,747	14,248	18,522	13,605
	Mexico	20,239	40,823	26,714	4,326	8,433	11,932	17,682	28,822
Total		36,241	55,408	42,835	15,809	25,180	26,180	36,204	42,427
Caribbean	Cuba								
Asia	Japan	219	1,246	1,544	20		5,014	877	1,466
	China-Taiwan	2,285			70	42	492	57	80
	South Korea	485	1,857		14,400				
	China-Mainland		497			57			
	Philippines								
	Indonesia		254						
	Malaysia	46							
	Vietnam								
	Thailand								62
Total		3,035	3,854	1,544	14,490	99	5,506	934	1,608
Middle East and North Africa	Turkey								
	Israel			9,096	28,269				2,786
	Egypt								
	Morocco								
Total				9,096	28,269				2,786
European Union	Denmark	447	2,968	35,664	72,584	137,720	120,627	139,877	115,719
	United Kingdom	9,432	7,014	58,570	59,638	119,327	86,588	86,584	87,489
	Ireland	36,645	64,454	179,295	166,471	156,027	146,708	218,828	241,743
	Netherlands	8,648	9,824	65,928	30,122	45,377	15,495	27,777	2,929
	France	3,836	6,994	71,907	43,952	36,203	31,990	35,864	24,447
	Germany	2,565			4,337		973		29,847
	Spain	6,636	6,257	104,436	73,163	36,902	52,890	78,110	72,311
	Portugal	19,983	20,352	113,669	117,549	85,221	90,160	79,833	97,122
	Italy		750	55,259	19,533	28,857	24,680	16,064	12,746
Total		88,192	118,613	684,728	587,349	645,634	570,111	682,937	684,353
Subtotal		127,468	177,965	738,203	645,917	670,913	601,797	720,075	731,174
Other countries		714	1,404	2,519	317	748	3,002	43	44,668
Total		128,182	179,369	740,722	646,234	671,661	604,799	720,118	775,842

Area	Importing country	2000/01	2001/02	2002/03	2003/04	Crop year 2004/05	2005/06	2006/07	2007/08	2008/09
		Metric tons								
North America	Canada	26,325	26,864	32,931	46,689	116,633	113,895	190,373	683,361	704,451
	Mexico	41,105	27,895	41,325	59,374	92,500	280,602	607,848	1,000,936	1,387,153
Total		67,430	54,759	74,256	106,063	209,133	394,497	798,221	1,684,297	2,091,604
Caribbean	Cuba						10,043	40,646	112,489	146,468
Asia	Japan	26	40	15		288	29,039	78,523	151,243	210,517
	China-Taiwan				3,732	29,227	73,126	126,272	172,530	181,393
	South Korea		14	50	106	2,111	11,497	87,730	175,515	279,622
	China-Mainland			18	42	54		1,150	4,593	97,936
	Philippines		51		377	5,374	49,210	67,349	95,896	127,618
	Indonesia				3,160	36,744	45,860	63,790	90,624	182,712
	Malaysia				7,097	24,614	33,682	32,141	53,687	47,009
	Vietnam				633	13,093	12,645	47,993	101,332	183,966
	Thailand			61	10	5,102	26,290	50,561	130,354	270,488
Total		26	105	144	15,157	116,607	281,349	555,509	975,774	1,581,261
Middle East and North Africa	Turkey					216	416	17,600	403,128	380,176
	Israel			6,032	6,348	45,694	8,607	40,256	167,379	176,982
	Egypt								31,308	67,843
	Morocco						20,254	37,318	68,705	74,450
Total				6,032	6,348	45,910	29,277	95,174	670,520	699,451
European Union	Denmark	140,497	130,287	88,856	20,539	117		660		
	United Kingdom	64,962	117,434	141,287	194,731	116,935	152,087	79,894	16,960	6,229
	Ireland	270,936	258,611	285,093	203,316	204,195	186,897	66,377	79,605	95,816
	Netherlands	30,326	10,295	15,943	35,109	54,344	9,014	18,690	18,846	
	France	5,659	3,000				1,121			
	Germany	31,442	30,920	580	5,144	26,213	9,749			
	Spain	30,699	85,310	37,380	61,184	90,836	68,455	26,267	49,698	14,999
	Portugal	76,896	67,498	61,315	58,547	68,848	50,038	8,054		
	Italy		22,032						384	
Total		651,417	725,387	630,454	578,570	561,488	477,361	199,942	165,493	117,044
Subtotal		718,873	780,251	710,886	706,138	933,138	1,192,527	1,689,492	3,608,573	4,635,828
Other countries		63,628	75,640	41,350	31,727	34,250	36,808	90,386	312,090	325,170
Total		782,501	855,891	752,236	737,865	967,388	1,229,335	1,779,878	3,920,663	4,960,998

Source: U.S. Department of Commerce (b), Bureau of the Census. http://www.usatrade online.gov/.

Table 8. Estimates of potential annual DDGS consumption, by livestock class

	Dhuyvetter et al. 2000-2004 average[1]		Berger and Good 2007[1]		Dooley 2008[1]		Fox 2008[1]		Average of all estimates	
	Thousand metric tons	*Percent*	*Thousand metric tons*	*Percent*	*Thousand metric tons*	*Percent*	*Thousand metric tons*	*Perce nt*	*Thousand metric tons*	*Percent*
Beef cattle	29,446	63.1	38,709	70.0	34,524	69.1	20,774	59.1	30,863	65.4
Beef cows	9,803	21.0	10,859	19.7	5,703	11.4	4,807	13.7	7,793	16.4
Cattle on feed	12,261	26.3	15,450	28.0	16,591	33.2	12,761	36.3	14,266	30.9
Other cattle	7,381	15.8	12,400	22.4	12,230	24.5	3,206	9.1	8,804	18.0
Dairy cattle	6,276	13.5	6,779	12.3	5,347	10.7	7,693	21.9	6,524	14.6
Dairy cows	6,276	13.5	6,779	12.3	5,347	10.7	7,693	21.9	6,524	14.6
Swine	3,663	7.9	3,824	6.9	3,842	7.7	3,677	10.5	3,752	8.2
Breeding swine	1,037	2.2	1,031	1.9	973	1.9	642	1.8	921	2.0
Market swine	2,626	5.6	2,793	5.1	2,869	5.7	3,035	8.6	2,831	6.3
Poultry	7,245	15.5	5,950	10.8	6,215	12.4	3,014	8.6	5,606	11.8
Broilers	4,486	9.6	4,263	7.7	3,709	7.4	1,809	5.1	3,567	7.5
Layers	1,818	3.9	1,686	3.1	1,644	3.3	683	1.9	1,458	3.0
Pullets	161	0.3			104	0.2			133	0.1
Turkeys	780	1.7			757	1.5	522	1.5	686	1.2
Total	46,630	100.0	55,261	100.0	49,929	100.0	35,158	100.0	46,744	100.0

Note: Totals may not add due to rounding. [1] Calendar year.

Source: Dhuyvetter et al., 2005; Berger and Good, 2007; Dooley, 2008; Fox 2008.

The imported Chinese DDGS price, relative to domestic Chinese corn prices or imported corn prices, appears to be the primary driver of DDGS imports into China for marketing year 2009/10. The availability of shipping containers, providing a "backhaul" opportunity, has facilitated the movement of this product relative to shipping in larger bulk ocean vessels.

Future consumption growth of DDGS is expected to moderate with a declining domestic share and rising export share (USDA, February 2010). The domestic share of the projected supply could range from 70 to 75 percent, with the export share ranging from 25 to 30 percent, compared with the 2009/10 estimated domestic share of 75 percent and export share of 25 percent.

Potential U.S. Feed Consumption Exceeds Future Supply of DDGS

Industry experts were concerned that the rapid acceleration of ethanol production could cause the supply of distillers' grains to exceed its potential use in the feed market. If this is the case, the feed market and other potential uses of DDGS, such as boiler fuel or soil amendments, could be affected. The authors of this report project that the future U.S. DDGS supply will be 38.6 mmt in 2019/20 based on USDA's Baseline Projections (USDA, February 2010). Selected estimates of current potential U.S. DDGS feed consumption range from 35.2 to 55.3 mmt (Dhuyvetter et al., 2005; Berger and Good, 2007; Dooley, 2008; Fox, 2008) (table 8). Thus, projected DDGS supplies do not appear to exceed the estimated average potential of 46.7 mmt for the domestic feed market.

Furthermore, potential exports of DDGS range from 20 mmt to 52 mmt. The U.S. Grains Council reports that the export market shows current potential use at 20 mmt (Keefe, 2008). Another estimate for potential export consumption comes in at 52 mmt. For example, Fox (2008) estimated 12 mmt for large international markets and Paulson (2008) estimated 40 mmt for small international markets. So, there appears to be sufficient use potential in both the U.S. and export feed market to consume the U.S. distillers' grain supply, assuming DDGS prices remain favorable for feeding.

Both domestic and export consumption estimates are derived from a fairly uniform method. Daily intake (as fed) per species is established and multiplied by the DDGS typical inclusion rate, multiplied by days fed per year, multiplied by the annual head (based on whether it is an annual inventory number or annual number produced) of selected beef cattle, dairy cattle, swine, and poultry. Summing for each type of livestock provides an estimation of the potential DDGS consumption level for the particular country in question. If any of these variables differ, then potential consumption estimates will also differ. For example, if inventory numbers rise or decline, the industry could generally expect potential DDGS consumption to respond accordingly. Estimates of daily intake and inclusion rates also differ by study, which accounts for some of the variance in potential consumption estimates. Also, some of the studies use a multi-year average estimate of livestock inventories, while others may use inventories from a specific year.

Potential consumption estimates derived from each of the referenced studies assume that DDGS were priced competitively and would enter the livestock/ poultry ration. There may be times when feed use of DDGS is curtailed or discontinued because of unfavorable prices for DDGS relative to competing ration ingredients. Also, there may be times when livestock/poultry producers switch from a typical inclusion rate of DDGS to a maximum inclusion rate because of more favorable DDGS prices relative to competing ingredient prices. Thus, changing the prices for livestock/poultry ration's competing ingredients could impact the estimated feed potential of DDGS.

In addition to market prices, the adoption of new ethanol/DDGS production technologies can influence the relationship between DDGS supply and its potential use. As the composition of DDGS (such as the removal of a portion of the oil) changes, then the use of DDGS for different types of livestock might change.

PRICE DISCOVERY, RISK MANAGEMENT, AND PRICE RELATIONSHIPS

The surge in ethanol production creates a question about the relationships between DDGS, corn, and soybean meal prices and how they may change.[22] Many grain prices are simultaneously determined by a futures market price and the cash market. Nearly half of the respondents to a recent survey stated that a distillers' grain futures contract would be an important risk management tool (Stroade et al., 2010).[23] A futures market for DDGS, however, did not exist until very recently.[24] We used a basic approach to analyze the price relationships among DDGS, corn, and soybean meal, indicating other areas of potential research as the market for DDGS develops.

Price Discovery and Risk Management

Very little scholarly research has been conducted on price discovery for the DDGS market. Stroade et al. attempted to address some of the price discovery and risk management questions for DDGS. Their survey results indicated that ethanol plants establish DDGS transaction prices based on corn futures prices (87 percent of respondents), soybean meal prices (43 percent of respondents), other sources (23 percent of the respondents), USDA published DDGS prices (16 percent of respondents), and no external sources (3 percent of the respondents). Slightly more than 40 percent of the respondents used multiple sources of information to price their DDGS.

The survey asked two questions about DDGS price risk management. The first question asked about the methods used to manage price risk management. A majority of the respondents used a price risk management method (64 percent of the respondents), while 36 percent stated they did nothing to manage price risk for DDGS. The second question asked about the methods used to manage DDGS price risk. For those using a price risk management method, corn futures prices (45 percent of the respondents) and forward contracts (40 percent of the respondents) were the price risk methods most reported. About 20 percent of the respondents used both corn futures and forward contracts. Only one respondent reported using soybean meal futures.

Schroeder (2009) found that DDGS prices are not effectively cross hedged based on traditional corn or soybean meal futures prices. Furthermore, Schroeder states that since there is no effective cross-hedging potential between DDGS prices and corn or soybean meal futures prices, a futures market for DDGS should be investigated. The corn distillers' dried grain futures contract, which began on April 26, 2010, is expected to bring much needed price discovery tools and price transparency to the market, but will also serve as a risk management tool for ethanol plants and livestock/poultry feeders. Actual trading of the contract over time will determine its success.

DDGS Price Relationship with Corn and Soybean Meal

Monthly DDGS prices (dry matter basis) for Central Illinois (fig. 1) ran at a premium to the corn price, with an average ratio of 1.09 for May 1995 to August 2006 (fig. 2). Beginning with marketing year 2006/07, however, monthly DDGS prices have averaged less than corn prices. The price ratio averaged 0.89 for September 2006 to August 2010 and ranged between 0.70 to

1.02. Thus, DDGS prices are falling into a range where they become more competitive as a feed source for many livestock and poultry rations.[25] The general decline in DDGS prices relative to corn (May 1995 to August 2006 compared with September 2006 to July 2010) is due, in part, to increased DDGS production. The increased seasonality within the marketing year also may be influencing the monthly price ratio. DDGS prices are higher in the winter months when pasture is not available, and then prices decline in the spring and summer as more pasture and forage become available. This is particularly interesting considering that there is an increase in DDGS consumption by beef cattle.

Monthly price ratios of DDGS to soybean meal (dry matter basis) averaged 0.49 for May 1995 to August 2006, compared with 0.44 for September 2006 to August 2010 (fig. 2). These price ratios are consistent with DDGS' status as a medium-protein feed, but the slight decline in the DDGS/soybean meal price ratio relative to DDGS/corn price ratio reflects DDGS' increased production and lower price. The increased DDGS supply, accompanied by a lower price, leads to a lower DDGS price relative to corn (usually less than corn). DDGS also substitutes for soybean meal, which may reduce the demand and price for soybean meal, but a sharper drop in the price of DDGS is reflected in a decline of the DDGS/soybean meal price ratio.

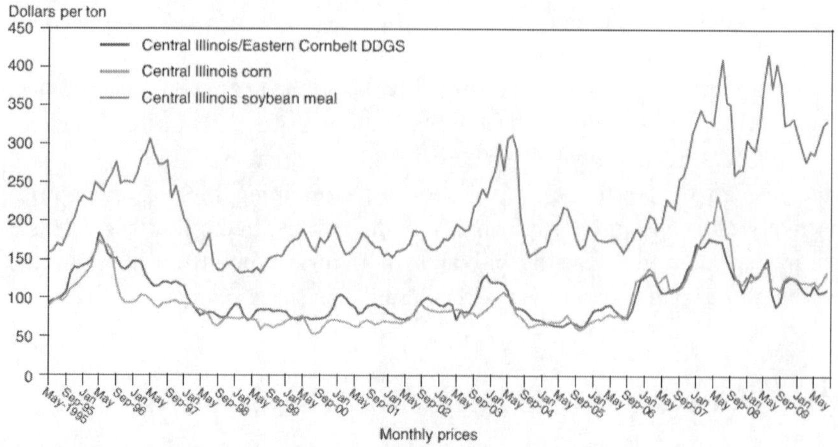

Note: Monthly prices for September-August marketing year.

Source: USDA, *Corn Belt Weekly Feedstuffs*; Missouri Department of Agriculture News Service, St. Joseph, Missouri, http://ams.usda.gov/mnreports/sj_gr225.txt; USDA, Economic Research Service, Feed Grains Database.

Figure 1. Monthly prices: Soybean, DDGS, and corn, May 1995-August 2010.

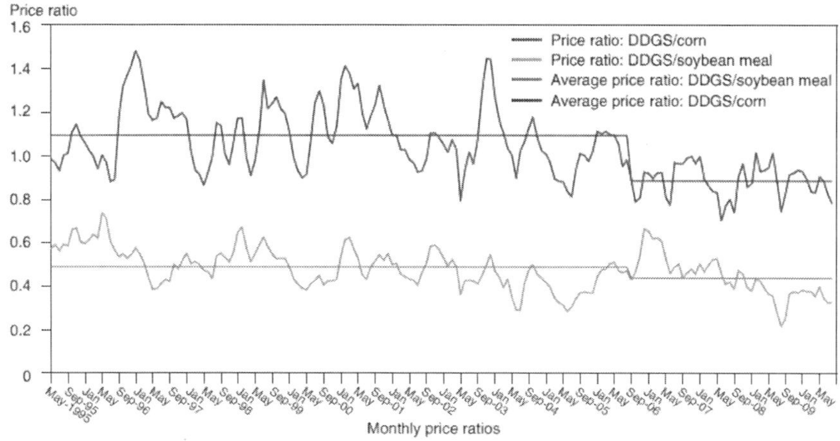

Note: Monthly prices for September-August marketing year.

Source: USDA, Corn Belt Weekly Feedstuffs; Missouri Department of Agriculture News Service, St. Joseph, Missouri, http://ams.usda.gov/mnreports/sj_gr225.txt; USDA, Economic Research Service, Feed Grains Database.

Figure 2. Monthly price ratios (dry matter basis): DDGS compared with corn and soybean meal, May 1995-August 2010.

A simple model of DDGS prices was estimated using two major feed components that are replaced by DDGS—corn (energy) and soybean meal (protein). Corn and soybean meal prices accounted for 98 percent of the variation in DDGS prices and thus are major determinants of the DDGS price (see equation 1 in box, "Major Determinants for the Price of DDGS are Corn and Soybean Meal Prices," p. 23). But the real issue is whether these relationships have changed with the surge in dry-mill ethanol production. We can now model how these relationships held with the surge in dry-mill ethanol production. As seen in equation 2 compared with equation 1, DDGS are increasingly used as an energy feed and, as a result, soybean meal prices may be starting to lose some of their impact on DDGS prices, as the coefficient value for soybean meal declined to less than half that reported in equation 1. The statistical significance of soybean meal prices as an explanatory variable for DDGS prices also declined slightly. An increase in cattle feeding, and, thus an increase in the seasonality of DDGS prices, may also contribute to the decline in the coefficient value for soybean meal prices. Further analysis should explore the effects of seasonality on this price relationship.

As reported earlier, DDGS are viewed as a mid-protein feed ingredient, but they are also rich in energy. As a result of these characteristics, the price of

DDGS usually runs above competing medium protein feeds, such as corn gluten feed, hominy feed, and wheat middlings (fig. 3).

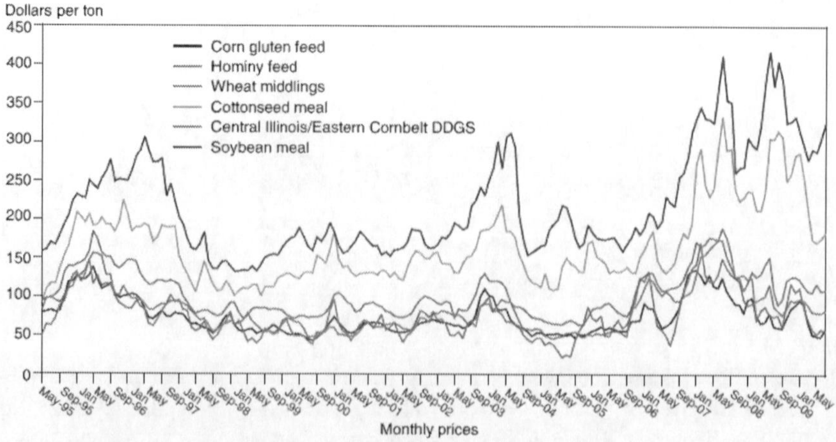

Note: Monthly prices for September-August marketing year.

Source: USDA, *Corn Belt Weekly Feedstuffs*; Missouri Department of Agriculture News Service, St. Joseph, Missouri, http://ams.usda.gov/mnreports/sj_gr225.txt; USDA, Economic Research Service, Feed Grains Database.

Figure 3. Monthly prices: DDGS and competing feed ingredients, May 1995-July 2010.

Regional DDGS Price Relationships

DDGS prices appear to be highly correlated among States and producers. The Pearson correlation coefficients of monthly DDGS prices among selected States (Central Illinois, Nebraska, Iowa, Minnesota, Kansas, and Missouri) for the most recent 3 marketing years (2006/07 through 2008/09) revealed that select State DDGS prices had a positive correlation with the coefficients near 0.90 or above (USDA, Corn Belt Weekly Feedstuffs).[26] DDGS price correlation coefficients for weekly DDGS prices over the same general time period at selected State ethanol plants as published by USDA's Agricultural Marketing Service (table 5) were also high, near 0.90 or above (USDA, AMS, *Bioenergy Market News* report). These findings suggest that DDGS prices across States and producers are highly correlated and move together. In contrast, Schroeder (2009) found that spatially separated DDGS markets do not have strong price relationships. His study focused, however, on the long term and included different markets with a time series beginning in 2001, when the market for DDGS was less mature. As DDGS production expands and DDGS are increasingly used as a substitute for corn and soybean meal in feed rations, prices should become more linked to corn and soybean meal prices. Corn and soybean meal prices are highly correlated among regions, with differences reflecting the cost of transportation and handling.

Increased feeding of beef cattle on the range and in the feedlot may increase DDGS demand during the late fall through winter months for Kansas and Nebraska as the DDGS to corn price ratio was highest during those months for 2007/08 and 2008/09 (fig. 4).[27] In the spring and summer, when pasture or forage is more available, DDGS prices tend to decline, exhibiting a degree of the seasonality mentioned earlier. DDGS are a new feed for many users and optimizing their use in rations is an ongoing effort.

Major Determinants for the Price of DDGS are Corn and Soybean Meal Prices

Central Illinois DDGS price is a function of Central Illinois corn and soybean meal prices (all prices were converted to a dry matter basis). Using monthly prices for May 1995 through August 2006 (dollars per short ton)—a period prior to the surge in dry-mill ethanol production—we derived the following regression equation:

$$\underset{(0.05)}{DDGSpr = 0.77\ cornpr} \quad \underset{(0.02)}{+0.14\ soymealpr} \qquad \text{Adj. } {}_{R}2\ 1 = 98\ \text{percent}^2$$

Where DDGSpr is the Central Illinois/Eastern Corn Belt price of DDGS, cornpr is the Central Illinois price of corn, and soymealpr is the Central Illinois price of soybean meal. Both corn and soybean meal prices are statistically significant at the 1 percent level.

Using monthly prices for September 2006 through July 2010 (dollars per short ton)—a period of increasing dry-mill ethanol production—we derived the following regression equation:

$$\underset{(0.05)}{DDGSpr = 0.77\ cornpr} \quad + \quad \underset{(0.03)}{0.05\ soymealpr} \quad \text{Adj. } R^2 = 97\ \text{percent}^2$$

Corn price is statistically significant at the 1 percent level, but the significance of the soybean meal price is at the 5 percent level.

1 The coefficient of determination, R^2, is the proportion of variability in a dataset (dependent variable) that is accounted for by the independent variables in the statistical model. It provides a measure of how well future outcomes are likely to be predicted by the model. R-squared always increases when a new variable is added to a model. Sometimes additional variables are added to a statistical model simply to increase the R-squared. To compensate for this tendency, the adjusted R-squared is used. The adjusted R-square is a modification of R-squared that adjusts for the number of variables in a model. Adjusted R-squared increases only if the new variable improves the model more than would be expected by chance (see Peter Kennedy, *A Guide to Econometrics,* 1992).

2 Numbers in parentheses below coefficient values are standard errors.

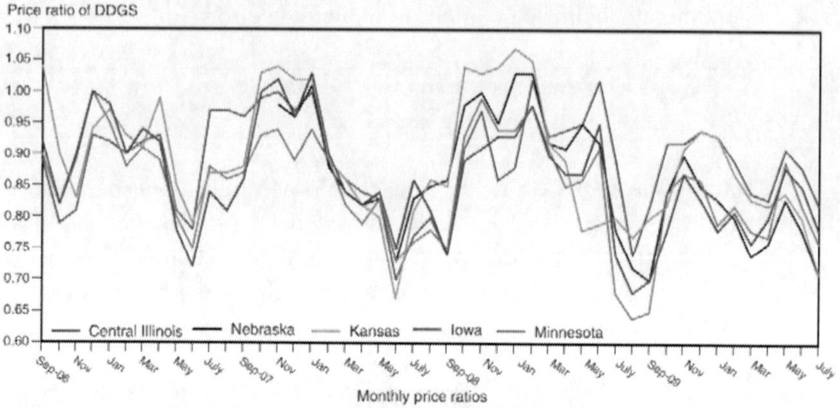

Note: Monthly prices for September-August marketing year.

Source: USDA, Corn Belt Weekly Feedstuffs; Missouri Department of Agriculture News Service, St. Joseph, Missouri, http://ams.usda.gov/mnreports/sj_gr225.txt; USDA, Economic Research Service, Feed Grains Database.

Figure 4. DDGS/corn price ratios (dry matter basis), by selected States, August 2006-July 2010.

Price advantages of feeding distillers' grains (dry, modified, and wet) relative to corn are reflected in figure 5 for Illinois and figure 6 for Nebraska.[28] For November 9, 2007 through September 3, 2010, weekly Illinois ethanol plant prices of dry, modified, and wet distillers' grains relative to corn averaged 84, 66, and 37 percent, respectively (all dry matter comparisons) (fig. 5). Nebraska weekly ethanol plant prices of dry, modified, and wet distillers' grains (November 9, 2007 through September 3, 2010) relative to corn were 87, 72, and 41 percent, respectively (all dry matter comparisons) (fig. 6).

Nebraska's price ratios were higher than for Illinois, likely due to Nebraska having a larger number of beef cattle relative to dairy cattle, hogs, or poultry.

The varying moisture content of distillers' grains could affect their potential in different markets in a particular State. Although WDGS had the lowest price of the three types of distillers' grains for Illinois and Nebraska, transportation, handling, and storage challenges seem to favor cattle feeding over swine and poultry. Wet and modified distillers' grains work best with beef-cattle rations, where higher percentages of the ration are possible.

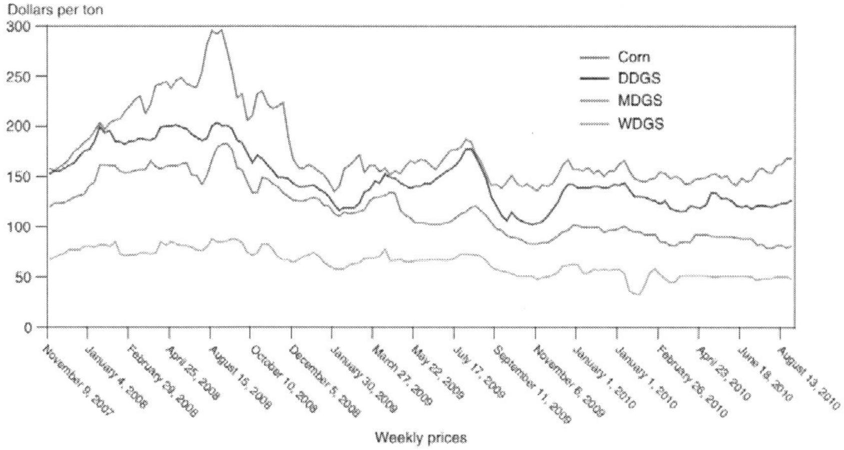

Note: Weekly prices for September-August marketing year.
Source: USDA, Agricultural Marketing Service, Bioenergy Market News reports.

Figure 5. Weekly prices of corn and dry, modified, and wet distillers' grains (dry matter basis): Illinois ethanol plant, November 9, 2007-September 3, 2010.

The average DDGS prices for Illinois and Nebraska were favorable for feeding livestock/poultry during this time period, particularly for feeders located near ethanol plants. While the modified and wet distillers' grain prices averaged less than the corn price, there were handling and storage costs associated with these higher moisture feeds that must be considered by feeders. Feeders must also consider transportation costs for each product and the slightly higher energy levels for WDGS in cattle rations. Despite these averages, there are periods when DDGS and MDGS prices are nearly equal to each other for Illinois and Nebraska. The difference in these prices could reflect the cost of drying. Schroeder (2009) suggests that if the dry, modified, and wet distillers' grain markets are closely linked and efficient, the price differences on a dry-matter basis across these product forms would essentially equal the cost of drying. The price behavior of natural gas would have to be examined to see if it influenced this relationship. Also, demand factors may permit higher prices for DDGS than the cost of drying would suggest.

Corn and soybean meal prices are important determinants of the DDGS price and may explain why some DDGS producers rely on corn and soybean meal futures prices to establish DDGS prices, and why some use corn and soybean meal futures as a risk management tool. Average DDGS prices now run at a discount to corn compared with an average premium prior to the surge in distillers' grain production. Our statistical analysis indicates that soybean meal prices are losing some of their explanatory power for DDGS prices, suggesting that DDGS are increasingly being used as an energy feed and substitute for corn. Further analysis will help explain the differences and behaviors of corn, DDGS, MDGS, and WDGS regional price relationships over time.

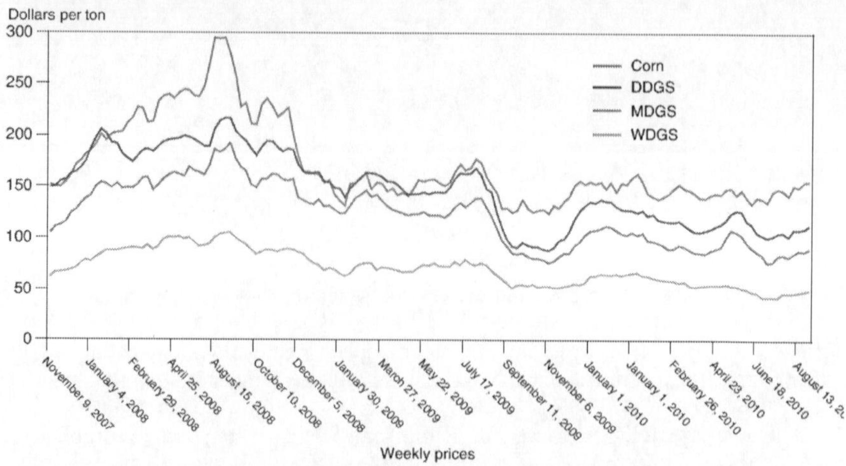

Note: Weekly prices for September-August marketing year.
Source: USDA, Agricultural Marketing Service, Bioenergy Market News reports.

Figure 6. Weekly prices of corn and dry, modified, and wet distillers' grains (dry matter basis): Nebraska ethanol plant, November 9, 2007-September 3, 2010.

SUMMARY AND CONCLUSION

The authors of this report present a methodology for estimating supply and consumption for the U.S. distillers' grain market, providing a transparency that contributes to the quality of the data, which can then be used to compare the supply, use, and price relationships of DDGS with other feed sources (see "Appendix: Estimation of U.S. Distillers' Grains Supply and Use"). Using these estimates, we show the size and composition of the distillers' grain market in broad terms. These estimates can then be used to develop a broader understanding of the DDGS and larger feed markets.

As new markets develop, buyers and sellers struggle to determine how best to position the product in the market and at what price. Our method of estimating DDGS supply and use was presented to develop estimates consistent with similar estimates for corn and soybean meal. Estimated supply of U.S. distillers' grains is expected to total 33.3 mmt for 2009/10, more than double that for 2006/07. Most of this additional supply comes from corn dry-mill fuel ethanol production, with a lesser amount coming from dry-mill beverage ethanol distilleries. Imports account for a minor segment of the market. Growth in both ethanol and distillers' grains are expected to moderate in the future. Changing dry-mill ethanol technology may alter the future composition of distillers' grains and the application of DDGS in livestock feed.

Domestic U.S. feed and residual use comprises three-fourths of U.S. distillers' grain disappearance. Domestic feeders have rapidly adopted this product to take advantage of its nutrient and cost advantages because it provides good value per unit of energy and protein. This adoption occurred early despite many issues of product quality, consistency, handling, and feed safety, due to education and common industry practices. Foreign feeders are also beginning to discover the economic feed value of distillers' grains, and the rate of export growth has surpassed that found in the domestic feed market. Exports represented about a fourth of U.S. distillers' grain disappearance in 2009/10.

The authors used the findings in this report to address questions about the potential size of the DDGS supply relative to its potential feed use. The potential for current DDGS consumption in the United States generally exceeds future U.S. supply estimates. Future supply of DDGS was estimated at 38.6 mmt for 2019/20 compared with an estimated current potential U.S. DDGS feed consumption of 35.2 to 55.3 mmt. Additionally, expected distillers' grain export potential is estimated to range from 20 to 52 mmt, which would cover any feed use not covered by U.S. needs.

Corn and soybean meal prices are important determinants of DDGS prices and may explain why some producers rely on corn and soybean meal futures prices to establish DDGS prices, or why some producers use corn and soybean meal futures as a risk management tool. Another DDGS price discovery and risk management tool has become available with the initiation of a futures contract on April 26, 2010 for distillers' grain at the CME Group. Average prices of DDGS for most the past 4 marketing years (2006/07- 2009/10) are now running at a discount to corn compared with an average premium prior to the surge (May 1995 through August 2006) in distillers' grain production. As DDGS increasingly became an energy feed, soybean meal prices lost some of their connection to DDGS prices. Perhaps the increased seasonality of DDGS prices could contribute to the explanation of this trend. Corn, DDGS, MDGS, and WDGS regional price relationships need further analysis to better explain their differences and behaviors over time.

REFERENCES

Ash, Mark S. *Animal Feeds Compendium*, Agricultural Economic Report Number 656, U.S. Department of Agriculture, Economic Research Service, May 1992.

Berger, Larry L., and Darrel L. Good. "Distillers Dried Grains Plus Solubles Utilization by Livestock and Poultry," *Corn-Based Ethanol in Illinois and the U.S.: A Report from the Department of Agricultural and Consumer Economics, University of Illinois*, Chapter 6, November 2007, http://www.farmdoc.illinois.edu/policy Ethanol%20Report.pdf.

CMEGroup. *Contract Specifications for Distillers' Dried Grain*, April 26, 2010, http://www.cmegroup.com/trading/agricultural/grain-and-oilseed distillers-dried-grain_contract_specifications.html.

CMEGroup. News release, February 16, 2010, http://cmegroup.mediaroom.com/index.php?s=43&item=2986&pagetemplate=article.

Cooper, G. "A Brief, Encouraging Look at "Theoretical" Distillers Grains Markets," *Distillers Grains Quarterly,* first quarter, 2006, http://ethanolproducer.com/dgq/article-print.jsp?article_id=1176.

Corn Refiners Association. *Corn Wet Milled Feed Products*, 4th edition, 2006, http://www.corn.org/wp-content/uploads/2009/12/Feed2006.pdf.

Coyle, William T. *Next-Generation Biofuels: Near-Term Challenges and Implications for Agriculture,* Outlook Report No. BIO-01-01, U.S. Department of Agriculture, Economic Research Service, May 2010, http://www.ers. usda.gov/Publications/BIO0101/.

Deutscher, Hope. "IDGC: Distillers Grains Products Markets Grow," *Ethanol Producer Magazine*, July 2009, http://www.ethanolproducer.com/article-print.jsp?article_id=5766.

Dhuyvetter, Kevin C., Terry L. Kastens, and Michael Boland. *The U.S. Ethanol Industry: Where Will It Be Located In The Future?* University of California, Agricultural Issues Center, November 2005, http://www.agmanager. info/agribus/energy

Dooley, Frank J. *U.S. Market Potential for Dried Distillers Grains with Solubles*, Working paper # 08-12, Purdue University, Department of Agricultural Economics, December 2008.

Fabiosa, Jacinto F., James Hansen, Holger Matthey, Suwen Pan, and Francis Tuan. *Assessing China's Potential Import Demand for Distillers Dried Grain: Implications for Grain Trade,* Staff Report 09-SR 104, Iowa State University, Center for Agricultural and Rural Development, December 2009, http://www.card.iastate.edu/publications/DBS/PDFFiles/09sr104.pdf.

Ferris, John N. *Modeling the U.S. Domestic Livestock Feed Sector in a Period of Rapidly Expanding By-Product Feed Supplies from Ethanol Production,* Staff Paper 2-006-4, Michigan State University, Department of Agricultural Economics, November 2006, http://ideas.repec.org/s/ags/midasp.html.

Food and Agricultural Food Policy Research Institute. Outlook reports, various issues, 2009, http://www.fapri.missouri.edu/outreach/

Fox, John A. "The Value of Distillers Dried Grains in Large International Markets," *Using Distillers Grains in the U.S. and International Livestock and Poultry Industries,* Iowa State University, Midwest Agribusiness Trade Research and Information Center at the Center for Agricultural and Rural Development, chapter 6, 2008, http://www.matric.iastate.edu/DGbook/distillers_grain_book.pdf.

Fox, John A. "Export Markets for Distillers Grains," *Distillers Grain Market Development & Price Risk Management Series FS3,* Kansas State University, Agricultural Experiment Station and Cooperative Extension Service, North American Institute for Beef Economic Research, August 2009, http://www.naiber.org/Publications/NAIBER/Export-Markets-for-Distillers-Grains.pdf.

Government Printing Office. *Energy Policy Act of 2005*, August 8, 2006, http://frwebgate.access bills&docid=f:h6enr.txt.pdf.

Government Printing Office. *Energy Independence and Security Act of 2007,* December 19, 2007, http://www.epa.gov/regulations

Keefe, Daniel. "International Trade in DDGS Supply and Demand Outlook, 2008-2015," presentation made at Economic Research Service conference on Emerging Issues in Global Meat Trade, Washington, DC, September 19, 2008.

Onetti, Silvia, and Eric Schwab. "New Ethanol Methods; New By-Product Feeds," *Hoard's Dairyman* Vol. 153, No. 15, September 10, 2008.

Paulson, Nicholas D. "International Demand for U.S. Distillers Dried Grains with Solubles in Small Markets," *Using Distillers Grains in the U.S. and International Livestock and Poultry Industries,* Iowa State University, Midwest Agribusiness Trade Research and Information Center at the Center for Agricultural and Rural Development, chapter 7, 2008, http://www.matric.iastate.edu/DGbook/distillers_grain_book.pdf.

Rendleman, C. Matthew, and Hosein Shapouri. *New Technologies in Ethanol Production,* Agricultural Economic Report Number 842, U.S. Department of Agriculture, Office of Energy Policy and New Uses, Office of the Chief Economist, February 2007, http://www.usda.gov/oce/reports/energy aer842_ethanol.pdf.

Renewable Fuels Association. *Changing the Climate: Ethanol Industry Outlook 2008*, p. 14, http://www.ethanolrfa.org/objects/pdf/outlook/RFA_Outlook_2008.pdf.

Renewable Fuels Association. *Feeding the Future*, September 2008, http://www.ethanolrfa.org/objects/documents/1914/feed_co-products.pdf.

Renewable Fuels Association. *2009 Ethanol Industry Outlook,* p. 24, http://www.ethanolrfa.org/objects/pdf/outlook/RFA_Outlook_2009.pdf.

Renewable Fuels Association. *Corn Quality Issues and Considerations for Ethanol Production,* November 2009, http://www.ethanolrfa.org/Grain%20Quality.pdf.

Renewable Fuels Association. *2010 Ethanol Industry Outlook*, pp. 6-7, http://www.ethanolrfa.org/pages/annual-industry

Renewable Fuels Association. *How Ethanol Is Made*, September 2010, http://www.ethanolrfa.org/pages/how-ethanol

Schingoethe, David J. "Distillers Grains for Dairy Cattle," presentation made at the Iowa Regional Distillers Grains Workshop, February 2004.

Schroeder, Ted C. "Distillers Grain Prices: Spatial Relationships, Arbitrage Opportunities and Risk Management," *Distillers Grain Market Development & Price Risk Management*, Series FS2, Kansas State University Agricultural Experiment Station and Cooperative Extension Service, North American Institute for Beef Economic Research, May 2009, http://www.naiber.org/Publications/NAIBER/Distillers_Grain_Prices.pdf.

Shurson, Jerry, and Abdorrahman S. Alghandi. "Quality and New Technologies to Create Corn Co-Products from Ethanol Production," *Using Distillers Grains in the U.S. and International Livestock and Poultry Industries,* Iowa State University, Midwest Agribusiness Trade Research and Information Center at the Center for Agricultural and Rural Development, chapter 10, 2008, http://www.matric.iastate.edu/DGbook/distillers_grain_book.pdf.

Staff, Charles, Executive Director, Distillers Grain Technology Council. Written communication, June 7, 2010.

Stroade, Jeri, Anikka Martin, Ann Conrad, and Ted Schroeder. "Distillers Grain Industry Price Discovery and Risk Management," *Distillers Grain Market Development & Price Risk Management*, Series FS4, Kansas State University, Agricultural Experiment Station and Cooperative Extension Service, North American Institute for Beef Economic Research, January 2010, http://www.naiber.org/Publications/NAIBER/Distillers-GrainPrice-Discovery&Risk-Management.pdf.

The ProExporter Network. *PRX Grain Database*, section I on ethanol, corn, and DDG, September 18, 2006 and October 2, 2007.

U.S. Department of Agriculture. *Corn Belt Weekly Feedstuffs*, Missouri Department of Agriculture News Service, St. Joseph, Missouri, http://www.ams.usda.gov/mnreports/sj_gr225.txt.

U.S. Department of Agriculture. "National Weekly Distillers Grains Summary," *Livestock and Grain Market News,* Des Moines, Iowa, http://www.ams.usda.gov/mnreports/lswndgs.pdf.

U.S. Department of Agriculture. "National Weekly Ethanol Summary," *Livestock and Grain Market News*, Des Moines, Iowa, http://www.ams.usda.gov/mnreports/lswethanol.pdf.

U.S. Department of Agriculture, Agricultural Marketing Service. *Bioenergy Market News* reports, http://www.ams.usda.gov/AMSv1.0/ams.fetchTemplateData.do?template=TemplateN&navID=MarketNewsAndTransportationData&leftNav=MarketNewsAndTransportationData&page=Bioenergy.

U.S. Department of Agriculture, Economic Research Service. Feed Grains Database, http://www.ers.usda.gov/data/feedgrains/.

U.S. Department of Agriculture, Office of the Chief Economist, World Agricultural Outlook Board (WAOB). *USDA Agricultural Projections to 2019,* Long-Term Projections Report OCE-2010-1, February 2010, http://www.ers.usda.gov/publications/oce101/.

U.S. Department of Agriculture, Office of the Chief Economist, World Agricultural Outlook Board (WAOB). *World Agricultural Supply and Demand*

Estimates, WASDE-444, March 12, 2010, http://www.usda.gov/oce/commodity/wasde/index.htm.

U.S. Department of Commerce (a), Census Bureau. *Fats and Oils: Oilseed Crushings Circular* M311J, table 2, http://www.census.gov/manufacturing

U.S. Department of Commerce (b), Census Bureau, http:www.usatradeonline.gov/.

U.S. Department of the Treasury, Alcohol and Tobacco Tax and Trade Bureau. Monthly statistical reports, 1992-2009, http://www.ttb.gov/statistics

U.S. Energy Information Administration. *Petroleum Supply Monthly,* February 26, 2010, http://www.eia.doe.gov/pub/oildata_publica tions/petroleum_supply_monthly/historical/2010/2010_02/ psm_2010_02.html.

U.S. Grains Council. *Biotech Traits Approved in EU Positive Move,* November 2009, http://grains.org/news-events/ 2015-biotech-traits-approved-in-eu-positive-move.

U.S. Grains Council. *DDGS Transport Proposal to the IMO Approved,* September 2010, http://www.grains.org.org/ global-update/2646-global-update-september-17-2010.

Voegele, Erin. "Going Global with DDGS," *Distillers Grains Quarterly,* first quarter, 2009, http://www.ethanolproducer.com/dgq/article-print.jsp?article_id=1280.

Watson, Stanley A. *Industrial Utilization of Corn*, American Society of Agronomy, pp. 721-63, 1977.

Weigel, Jerry C., Dan Loy, and Lee Kilmer. *Feed Co-Products of the Corn Wet Milling Process: Featuring Corn Gluten Feed and Corn Gluten Meal,* ExSeed Genetics and Iowa State University in cooperation with Iowa Corn Promotion Board, Renewable Fuels Association, and U.S. Grains Council, April 2004, http://www.extension.iastate.edu/Bioeconomy/Livestock.

Weigel, Jerry C., Dan Loy, and Lee Kilmer. *Feed Co-Products of the Dry Milling Process: Featuring Distillers' Dried Grains*, ExSeed Genetics and Iowa State University in cooperation with Iowa Corn Promotion Board, Renewable Fuels Association, and U.S. Grains Council, April 2004, http://www.extension.iastate.edu/Bioeconomy/Livestock.

Weiss, Bill, Maurice Eastridge, Dianne Shoemaker, and Normand St-Pierre. "Distillers Grains," *Ohio State University Extension Fact Sheet,* June 2007, http://ohioline.osu.edu/as-fact/pdf/distillers.pdf.

Wisner, Robert. "Understanding the Distillers Grains Supply/Usage Balance Sheet," *AgMRC Renewable Energy Newsletter,* November/December

2008, http://www.agmrc.org/renewable_energy the_distillers_grains_ supplyusage_balance_sheet.cfm.

Wisner, Robert. *Estimated U.S. Dried Distillers Grains with Solubles (DDGS) Production and Use,* Iowa State University, University Extension, http://www.extension.iastate.edu/agdm/crops. pdf.

APPENDIX: ESTIMATION OF U.S. DISTILLERS' GRAIN SUPPLY AND USE: A TRANSPARENT METHODOLOGY TO ESTIMATE U.S. DISTILLERS' GRAIN PRODUCTION AND CONSUMPTION

The need for a transparent method to estimate the supply (production and imports) and disappearance (domestic feed consumption and exports) of distillers' grains is apparent because of the number of undocumented estimates of production and consumption that have been made and discrepancies and inconsistencies among these estimates (Voegele, 2009; Food and Agricultural Policy Research Institute (FAPRI), 2009; RFA, 2008; The ProExporter Network, 2007). Our methodology is similar to the process used by USDA's Economic Research Service to estimate supply and use for the major livestock feed sources, which distillers' grains have rapidly become. A consistent methodology allows analysts to research issues with distillers' grains, such as supply and disappearance (domestic and export). Our estimation method relies on published data, but inventories are not included in our estimate of supply, since there are no published data for inventories of distillers' grains.

Production estimates for DDGS have become available, but these estimation methods are less than transparent and difficult to apply generally (Wisner, 2008; FAPRI; Deutscher, 2009; RFA, September 2008; U.S. Department of Commerce, Census Bureau (a)).[1] U.S. Census Bureau statistics are viewed by Government and industry analysts as incomplete and under-report DDGS production based on other reports of ethanol production. For example some DDGS production estimates are expressed on a calendar year basis, while other estimates are for a crop marketing year; assumptions about conversion factors and product breakouts also likely vary. For example, Wisner (2008) estimated production and exports of distillers' grain in 2007/08 at 25.7 and 3.9 million short tons, respectively. FAPRI estimated production and exports of DDGS in 2007/08 at 22.8 and 2.4 million short tons, respectively. RFA reported that 14.6 mmt of DDGS were produced in calendar year 2007 (RFA, 2008). In a later document, RFA reports estimated production of distillers' grains at 19.3 million metric tons for corn marketing year 2007/08 (RFA, September 2008).

Differences in estimates can occur due to assumptions on yield of distillers' grains per bushel of corn turned into ethanol. Estimates range from 16.75 to 18 pounds per bushel of corn fermented for ethanol. Some analysts assume that all corn used for ethanol production is credited to the dry-mill process. Dry mills account for an increasing share of total ethanol production, but this process is not used across all ethanol production.

In May 2007, the Census Bureau began publishing information that provided the quantity of corn grind used to produce ethanol along with its feed coproducts of corn gluten meal, corn gluten feed, wet corn gluten feed, and corn germ meal (U.S. Department of Commerce, Census Bureau (a)). The corn and other grain grind from the dry mills, including beverage distillation, were provided along with the feed coproducts—distillers' wet grain, distillers' dried grain with solubles, distillers' dried grains, distillers' dried soluble, and condensed distillers' solubles. A transparent methodology to estimate the production of distillers' grains allows for cross checking with the production estimates provided by the Census Bureau as well as with other industry or university production estimates. Census Bureau production estimates do not specify the moisture content of their listed coproducts. For example, university and USDA/AMS report moisture content of distillers' dry grains with solubles at 10 to 12 percent, modified wet distillers grains with solubles at 50 to 55 percent moisture, and wet distillers' grains with solubles at 65 to 70 percent moisture. Thus, it is difficult to determine whether U.S. Census Bureau's production estimates are similar to production estimates from other sources. Furthermore, our methodology can estimate data prior to the start of the Census Bureau statistics in May 2007.

Currently, the Census Bureau provides estimates for U.S. soybean meal grind and U.S. wheat flour grind that are used by both the Government and the industry. With time, the Census Bureau estimates of feedstocks used for both wet- and dry-mill ethanol plants and their resulting ethanol feed coproducts should become more reliable and useful to the distillers' grain market to gauge supply and consumption.

Meanwhile, we present a methodology that produces a transparent source of information on feedstocks for both wet- and dry-mill ethanol plants and their resulting feed coproducts that will be useful for the distillers' grain market to gauge supply and consumption. We followed USDA's methodology to estimate supply and demand for commodities:

- For production, we estimated the amount of corn used by beverage distilleries and the resulting amount of distillers' grains produced from this production process.
- Next, we estimated the amount of corn used by dry-mill ethanol plants and the resulting distillers' grains.
- Then we added imports of distillers' grains derived from U.S. Department of Commerce trade data and from the Economic Research Service's Feed Grains Database (USDA, ERS).
- By adding these components, we estimated supply of distillers' grains.
- From this supply, we subtracted exports derived from U.S. Department of Commerce trade data and from the ERS Feed Grains Database to arrive at feed and residual use.

Please note that the split for ethanol production between wet- and dry-mill corn ethanol plants is an estimate. Not all relevant data were available to make these estimates, but the estimation steps follow. For example, the corn oil yield per crop year should vary by crop year following actual oil yield variation by crop year. Since these data were unavailable, a constant estimate of 1.64 pounds per bushel processed was used. Furthermore, the amount of corn oil produced by dry mills later becomes an issue once fractionation began at some of the dry-mill plants. The Census Bureau report (U.S. Department of Commerce, Census Bureau (a)) may not be picking up this increase in oil from dry mills, so an assumption was made to account for this situation. Should actual data for estimates of fuel ethanol produced by wet mills become available, these data should be used instead of our estimates.

Estimated Supply

Production

Distillers' grains are produced from two sources—dry-mill beverage distilleries and dry-mill fuel ethanol producers. An estimation procedure for the amount of distillers' grains produced is explained below.

Dry-mill beverage distilleries—Our goal was to estimate the amount of corn used by dry-mill beverage distilleries and the amount of distillers' grains produced from this production process. The quantity of distillers' grains produced is shown in table 1, column 1. The amount of corn used was based on U.S. Department of the Treasury, Alcohol and Tobacco Tax and Trade

Bureau data. Bushels of corn used for distilled spirits were multiplied by 17.5 pounds to arrive at the amount of distillers' grains produced per bushel of corn used. As observed from table 1, this production figure has remained rather stable in the past few years at 0.9 mmt, but the share of total corn used for ethanol production has declined.

Dry-mill fuel ethanol—We needed to estimate the amount of corn used by dry-mill fuel ethanol plants. Since we know the total amount of corn used for fuel ethanol from the monthly World Agricultural Supply and Demand Estimates (WASDE), we had to determine the amount of corn used by wet-mills and subtract it from the total to arrive at the amount of corn used by dry-mills for fuel ethanol production. The reason for this separation is that different coproducts result from each production method.

Estimation methodology—The quantity of corn oil produced by wet corn mills must first be converted into the quantity of corn processed by wet mills. Corn oil is a coproduct created by wet mills, regardless of the other products produced, such as high-fructose corn syrup (HFCS), glucose/dextrose, starch, or ethanol. As seen in estimation step 1 (table 2, column 1), the total amount of corn processed by corn wet mills can be computed from total corn oil production minus that produced from dry mills divided by the yield of oil per wet-mill bushel processed. Both corn oil numbers produced at wet and dry mills were taken from the U.S. Department of Commerce (a), Census Bureau data. In the past 3 years, the corn oil amount produced by dry mills was increased to account for oil produced from fractionation (front-end), increasing this estimate from 1.9 percent for marketing year 2006/07, to 2.0 percent for marketing year 2007/08, and to 2.25 percent for marketing years 2008/09, 2009/10, and 2010/11.[2] Next, the amount of corn used for fuel ethanol production by wet mills must be determined. Since we estimated total corn used by the wet-mill plants, we subtracted the amount of corn used for HFCS, glucose/dextrose, and starch to arrive at the corn available for ethanol production as shown in estimation step 2 (table 2, columns 3-8).

Estimation steps

1. Total corn oil produced for 2009/10 was estimated at 2.50 billion pounds (table 2, column 1) minus 2.5 percent of the total estimated to be produced by dry mills equaled 2.44 billion pounds of corn oil produced by wet mills (table 2, column 2) divided by 1.64 pounds of corn oil per wet mill bushel processed equaled 1,490.1 million bushels of corn processed by corn wet mills (table 2, column 3).

2. Total bushels of corn estimated to have been processed by wet-mill corn plants in 2009/10 equaled 1,490.1 million bushels (table 2, column 3) minus corn used to produce HFCS equal to 515.0 million bushels (table 2, column 4) plus glucose/dextrose equaled 255.0 million bushels (table 2, column 5) plus starch equaled 245.0 million bushels (table 2, column 6) equaled 475.1 million bushels of corn used by wet-mill plants to produce fuel ethanol (table 2, column 8).

3. Total corn estimated to be used for fuel ethanol production in 2009/10 equaled 4,535.0 million bushels (table 3, column 1) minus corn used by wet mills for fuel ethanol production which equaled 475.1 million bushels (table 3, column 2) equaled 4,059.9 million bushels of corn used by dry mills for fuel ethanol production (table 3, column 3).

4. 4,059.9 million bushels of corn used by dry mills for ethanol production (table 3, column 3) multiplied by 17.4 pounds of DDGS per bushel of corn processed by dry-mill ethanol plants equaled 32.0 mmt produced (pounds divided by 2204.622 pounds per metric ton) (table 1, column 2).

[1] A corn oil yield of 1.64 pounds per bushel is used for these calculations. This yield was computed as follows: 15.5 percent moisture content per bushel, test weight of 56 pounds per bushel and a fat content per bushel of 4.3 percent of dry matter (.043 x 47.32 lbs/bu. = 2.035 lbs. /bu.). Next, we assume that 85 percent of the oil is located in the germ and an extraction rate of 95 percent, yielding (2.035 lbs/bu. x .85x .95 = 1.64 pounds/bu.) (Corn Refiners Association, p. 5, 2006).

The bushels of corn used by wet mills for fuel ethanol produced from the total corn used by both wet and dry mills for fuel ethanol production must then be subtracted. Total corn used by both wet and dry corn mills for fuel ethanol production can be obtained from USDA's *WASDE*. The difference is the amount of corn used by dry mills for fuel ethanol production (see table 3). This calculation is illustrated in estimation step 3.

The final step, as seen in estimation step 4, is to multiply the amount of corn used by dry mills to produce ethanol by 17.5 pounds—the weight of distillers' grains after the fermentation process (table 1, column 2). Common estimates for this weight range from 16.75 pounds to 18 pounds of DDGS per bushel of corn fermented for ethanol production. We used 17.5 pounds for the earlier years and 17.4 pounds for the later years 2007/08 through 2010/11 because of fractionation, both front-end and back-end, and the removal of corn oil during the production of fuel ethanol by the dry mills.[3]

Since 2002/03, most U.S. ethanol was produced from dry-mill corn ethanol plants. The share of corn use by dry mills continues to rise, ranging from about 54 percent in 2002/03 to 90 percent in 2009/10 (see table 3).

Imports

Data on imports of distillers' grains were obtained from the U.S. Department of Commerce (b) and the Economic Research Service's Feed Grains Database. Unfortunately, distillers' grains are part of a category called "brewers and distillers dregs and wastes." Brewer's grains are found in this category, but so too are noncorn distillers' grains. Currently, most of these imports originate from Canada and could include noncorn feedstocks, such as barley or wheat. Regardless of the unknown products in this category, adjustments seem unwarranted since the quantity is small, ranging from 0.1 to 0.5 mmt (table 1, column 3).

Estimated Use

Domestic Feed and Residual Use

Residual use was computed by subtracting exports from supply (table 1, column 4). The quantity of U.S. distillers' grain supply for domestic use in 2009/10 was estimated at 25.0 mmt (table 1, column 1). Domestic feed and residual use accounted for 75 percent of use, but this share is declining.

Exports—Data on exports of distillers' grains were obtained from the USDA Feed Grains Database. Despite the title of this export category—brewers' and distillers' dregs and wastes—most of the product in this category comes from distillers' grains. For example, U.S. production of brewers' grains for 1988/89 was 117,900 short tons, the last year of published data (Ash, 1992). If one were to assume that 18 percent of this production was still being exported, the portion of production exported in 1988/89, we would expect that 117,900 short tons x 18 percent = 21,222 short tons or 21,222 short tons x 0.9071 = 19,252 metric tons would be exported. Regardless of the unknown product in this category, adjustments seem unwarranted since the quantity is small. Exports of 19.3 metric tons of brewers' grains would be considered high since corn consumption for beer production has declined since 1988/89. Exports of distillers' grains have accelerated since 2003/04, from less than 1.0 mmt to an estimated 8.3 mmt for 2009/10 (table 1, column 5).[4] Accelerating exports are expected to reach 25 percent of total disappearance in 2009/10, despite increasing U.S. feed and residual use of distillers' grains.

End Notes

[1] In addition to Government policies and longer run energy prices, the growth rate of the ethanol market depends, in part, on feedstock costs, changes in technology, and changes in Government incentives.

[2] The term "distillers' grains" refers to coproducts generated by dry-mill ethanol plants, including distillers' wet grains (DWG), distillers' dried grains (DDG), distillers' wet grains with solubles (DWGS), distillers' dried grains with solubles (DDGS), and condensed distillers' solubles (CDS). Unless otherwise specified for the remainder of this report, the term distillers' grains will mean distillers' dried grains with solubles (DDGS).

[3] Ethanol production numbers are expressed in terms of corn's September-August marketing year (U.S. Energy Information Administration, 2010).

[4] Most of these imports originated in Canada and their feedstock is unknown, perhaps consisting of wheat, barley, or corn. Since the quantity is small, the concern over the feedstock is less important.

[5] Brewers' grains, a byproduct of beer manufacturing, also has a long history of use in the livestock feed market. Currently, the quantities of these beer-manufacturing byproducts are dwarfed in comparison with the quantity of distillers' spent grains being produced. Most of the brewers' grains are reportedly fed in the wet form to nearby livestock feeders (Ash, p. 91).

[6] These fractionation technologies will be explained later in this report. See p.12, "Composition of Distillers' Grains is Beginning to Change."

[7] Currently, USDA reports (Feed Grains Database, yearbook table 29) quantities of corn gluten feed and corn gluten meal that are produced when making starch, high fructose corn syrup, and dextrose. Estimates of corn gluten feed and corn gluten meal that are produced from wet-mill ethanol production were unavailable and thus not included in the quantities reported in table 29. Consequently, the quantity of wet-mill feed coproducts is under-reported. New estimates of corn gluten feed and corn gluten meal produced from corn wet mills are made to account for this under-reporting and provided later in this report.

[8] For more information on the wet corn milling process and the feed coproducts produced, see *Corn Wet Milled Feed Products*, pp. 6-11, http://www.corn.org/wp-content/uploads/2009/12/Feed2006.pdf, (Corn Refiners Association, 2006); *How Ethanol Is Made* at http://www.ethanolrfa.org/resource/made/ (Renewable Fuels Association, 2010); or *Feed Co-Products of the Corn Wet Milling Process* at http://www.extension.iastate.edu/Bioeconomy/Livestock/ (Weigle et al., 2004).

[9] "As fed" or "as is basis" means that there is no adjustment for moisture content in comparison with dry matter basis where adjustments are made for moisture content.

[10] For more information on the dry mill process and the feed coproducts produced, see *How Ethanol Is Made* at http://www.ethanolrfa.org/resource/ made/ (Renewable Fuels Association, 2010) or *Feed Co-Products of the Dry Corn Milling Process* at http://www.extension.iastate.edu/Bioeconomy/ Livestock/ (Weigel et al., 2004).

[11] The production estimates for corn gluten feed and corn gluten meal reflected in table 4 compute the total produced from making HFCS, glucose/dextrose, starch, and ethanol.

Existing ERS estimates from table 29 compute corn gluten feed and corn gluten meal from only the production of HFCS, glucose/dextrose, and starch (USDA, ERS, Feed Grains Database, yearbook table 29, 2010).

[12] Similar concerns can be expressed about the U.S. DDGS market and may account for much of the decline in U.S. sales of DDGS to Europe, a topic addressed later in this report.

[13] Some plants already capture the carbon dioxide and others have started. One challenge for the industry is finding a market for the carbon dioxide collected (Rendleman and Shapouri, 2007).

[14] Currently, there are only two or three plants using front-end fractionation due to costs and complications from production and marketing multiple coproducts (Staff, 2010).

[15] A monogastric organism has a simple single-chambered stomach, whereas ruminants have a complex four-chambered stomach. Examples of monogastric animals include pigs, dogs, and cats. Poultry also have a monogastric digestive system.

[16] Smaller gains for corn-based ethanol are projected over the next 10 years and reflect only moderate growth in overall gasoline consumption in the United States, limited potential for further market penetration of ethanol into the E10 (10-percent ethanol blend) market (the blend wall), and the small size of the E85 (85-percent ethanol blend) market. In the latter years of the projections, production of fuel ethanol accounts for 34-35 percent of total corn use, and corn-based ethanol production exceeds 9 percent of annual gasoline consumption.

[17] Assumes 1 mmt of DDGS comes from beverage distilleries, 0.5 mmt is imported and ((5.025 billion bushels x 93 percent dry-mill grind) x 17.5 pounds of DDGS per bushel of corn processed) / 2204.622 pounds per metric ton = 38.6 mmt supply.

[18] If we assume that 15 billion gallons of corn-based ethanol will be produced in 2019/20, corn use for this dry-mill fuel ethanol could total 5.17 billion bushels (2.7 gallons ethanol per bushel of corn and 93 percent of ethanol produced from dry-mill plants), which would produce about 41.0 mmt plus 1 mmt from beverage distilleries, and .5 mmt from imports = about 42.5 mmt supply.

[19] The U.S. Grains Council supports an active research/demonstration program of the benefits of feeding distillers' grains and reports success in many of its demonstration countries. Furthermore, they estimate a potential 20 mmt of DDGS exports from the United States, indicating significant growth potential.

[20] Despite some insurance companies insisting that DDGS be treated as hazardous cargo, DDGS exports have grown rapidly in the past several years. As a result of DDGS' status as hazardous cargo, there have been some shipping disruptions and higher transport costs for DDGS exports. The U.S. Grains Council was instrumental in coordinating an industry effort to define DDGS as nonhazardous cargo. These efforts, in cooperation with the U.S. Coast Guard led to a U.S. proposal that was submitted to the International Maritime Organization (IMO). A subcommittee of the IMO approved the U.S. proposal to classify DDGS as nonhazardous cargo, but the proposal must be ratified by the IMO's Maritime Safety Committee. Industry experts expect this to occur during the committee's December 2010 meeting (U.S. Grains Council, 2010).

[21] Although the EU could increase its imports of DDGS with the approval of three types of genetically enhanced corn for food, feed, import, and processing as of October 30, 2009, this potential could be replaced with DDGS from fuel ethanol plants being built in Great Britain, Europe, and Africa (U.S. Grains Council, 2009).

[22] Initially, a useable price series was a problem, but a Central Illinois (later referred to as Eastern Corn Belt) price series was found and used (USDA, *Corn Belt Weekly Feedstuffs*, 2010). These data are available for every Tuesday from May 1995 to the present. Later, USDA's

Agricultural Marketing Service began providing weekly DDGS prices for different markets, such as Iowa, Wisconsin, Minnesota, Missouri, and Nebraska. USDA's Agricultural Marketing Service also began providing ethanol plant reports that covered the price of corn, distillers' grains (DDGS, MDGS, and WDGS), and ethanol (USDA, AMS, *Bioenergy Market News* report, 2010; USDA, "National Weekly Distillers' Grain Summary," 2010; USDA, "National Weekly Ethanol Summay," 2010) Corn and soybean meal prices for Central Illinois were obtained from USDA, ERS, Feed Grains Database.

[23] Information presented in the price discovery and price risk management sections are based on a 2010 survey by Stroade, Martin, Conrad, and Schroeder of ethanol plants listed by the Renewable Fuels Association.

[24] On February 16, 2010, the CME Group announced a new futures contract for DDGS. Trading began on April 26, 2010 and contract specification information can be found at http:// www.cmegroup.com.

[25] Figure 2 highlights the fact that the first time period is much longer than the more recent period characterized by the surge in distillers' grain production. Thus, a degree of caution should be used while interpreting these differences.

[26] In statistics, the Pearson product-moment correlation coefficient (referred to as PMCC and typically denoted by r) is a measure of the correlation (linear dependence) between two variables X and Y, giving a value between +1 and −1 inclusive. PMCC is widely used in the sciences as a measure of the strength of linear dependence between two variables.

[27] Corn prices are seasonally low at harvest, contributing to the higher ratio in the fall months.

[28] Prices are based on those reported at ethanol plants.

End Notes for Appendix

[1] The U.S. Census Bureau began publishing production estimates of distillers' grains in May 2007, but this estimating procedure does not yet appear to capture all production based on other less direct methods, including the methodology examined in this report. The Renewable Fuels Association released a study in September 2008 that contained a production estimate of DDGS, corn gluten feed, and corn gluten meal for 2007.

[2] Future adjustments must be based on whether corn oil continues to be removed from the fractionation (front-end) process.

[3] The reduction of distillers' grain yield per bushel of corn processed for dry-mill ethanol production from 17.5 to 17.4 is an attempt to capture the effects of fractionation and removal of oil from ethanol production process by dry-mill plants. Additional analysis will be required to derive a more precise estimate of these effects.

[4] Similar calculations were made by Dooley in December 2008, but were based on RFA calendar year production numbers and calendar year exports.

In: Distillers' Dried Grains
Editor: Justin N. Schuber
ISBN 978-1-62417-683-8

Chapter 2

Estimating the Substitution of Distillers' Grains for Corn and Soybean Meal in the U.S. Feed Complex*

Linwood A. Hoffman and Allen Baker#

Abstract

Corn-based dry-mill ethanol production and that of its coproducts—notably distillers' dried grains with solubles (DDGS)—has surged in the past several years. The U.S. feed industry has focused on the size of this new feed source and its impact on the U.S. feed market, particularly the degree that DDGS substitute for corn and soybean meal in livestock/poultry diets and reduce ethanol's impact on the feed market. This study develops a method to estimate the potential use of U.S. DDGS and its substitutability for corn and soybean meal in U.S. feed rations. Findings demonstrate that, in aggregate (including major types of livestock/poultry), a metric ton of DDGS can replace, on average, 1.22 metric tons of feed consisting of corn and soybean meal in the United States. Over time, DDGS may substitute for less corn and more soybean

* This is an edited, reformatted and augmented version of United States Department of Agriculture, A Report from the Economic Research Service, Publication No. FDS-11-I-01, dated October 2011.

Allen Baker is a retired agricultural economist from USDA, Economic Research Service.

meal as the share of beef cattle consumption of DDGS declines slightly (although increasing in absolute terms), with offsetting share increases in dairy cattle, swine, and poultry. Feed market impacts of increased corn use for ethanol are smaller than that indicated by the total amount of corn used for ethanol production because of DDGS.

INTRODUCTION

U.S. ethanol production growth has been stimulated partly by higher energy prices and the influence of the Energy Policy Act of 2005 and the Energy Independence and Security Act of 2007 (Government Printing Office, 2006 and 2007).[1] The 2007 Energy Act mandates the use of as much as 15 billion gallons of starch-based ethanol (mostly from corn) in the United States by 2015. In 2005/06, 4.5 billion gallons of ethanol were produced. By 2008/09, production more than doubled to 10.2 billion gallons, and in 2009/10, it rose to 12.5 billion gallons.2, [3] Accompanying this growth in ethanol production, distillers' grain production has expanded substantially, increasing from an estimated 10.4 million metric tons (mmt) in 2005/06 to an estimated 33.4 mmt in 2009/10.[4]

Distillers' dried grains with solubles (DDGS) have long been substituted for corn and soybean meal in livestock and poultry diets, supplying both energy and protein.[5] Incorporating DDGS into these diets, however, may be constrained by nutritional and/or price considerations. For example, DDGS may not provide some of the nutrients needed by a particular type of livestock or poultry to achieve its desired weight gain or maintain its body weight. Assuming DDGS provide the desired nutrients, the price of DDGS must be favorable for feeding. Hoffman and Baker (2010) show that DDGS prices relative to alternative feed ingredients have declined into a favorable feeding range for many different types of livestock and poultry.[6]

How does the continued market expansion of DDGS affect the U.S. feed complex (or processed feeds fed) and to what extent do DDGS substitute for corn and soybean meal, thereby reducing the grain market impact of using corn to produce ethanol? A bushel of corn used for dry-mill ethanol production yields DDGS equal to about a third of the corn's original weight. Researchers are less sure, however, how this coproduct substitutes for corn and soybean meal in the diets of different types of livestock and poultry. Understanding of the characteristics of DDGS may demonstrate more effectively how they can be substituted into different livestock/poultry diets and the impact this substitution may have on the U.S. feed complex. To compute the overall or aggregate DDGS substitution rate for corn and soybean meal, multiply each type of livestock/poultry's DDGS substitution rate for corn (energy) and soybean meal (protein) times the market share of DDGS consumption by type of livestock/poultry and then sum each of the products.

This report provides a transparent method to estimate the substitution potential of DDGS for corn (energy) and soybean meal (protein) and the corresponding impact this has upon the U.S. feed complex. First, the feeding characteristics of DDGS are reviewed along with potential inclusion rates for each type of livestock/poultry. Second, potential U.S. DDGS inclusion rates per livestock/poultry are determined. Third, substitution rates of DDGS for corn and soybean meal are determined by type of livestock/poultry. Fourth, DDGS consumption estimates (market share) by crop year are estimated by type of livestock/poultry. Fifth, the aggregate substitution of DDGS for corn and soybean meal is computed by multiplying the market share times the substitution rates between DDGS and corn and soybean meal by type of livestock/poultry. Lastly, impacts on the U.S. feed complex are determined from the substitution of DDGS for corn and soybean meal.

Feeding DDGS to Livestock/Poultry

Feeders that choose to include DDGS into the diets of livestock/poultry need to be aware of DDGS nutritional content and feeding issues related to the use of these nutrients. The amount of DDGS that can be included in the diet of a particular type of livestock/poultry varies by its nutrient requirements and the nutrient availability and cost of alternative diet ingredients.

DDGS Nutritional Content Issues

DDGS are used by the livestock and poultry industries as a source of protein and energy in feed rations. DDGS are considered a mid-protein feed that offers the same or greater energy as corn but contains less protein than soybean meal (table 1). Ruminant animals, such as beef and dairy cattle, can use distillers' grains nutrients more readily than monogastric animals, such as hogs and poultry.[7] Compared with corn, DDGS are higher in calcium, phosphorus, and sulfur (table 1) so that, depending on the inclusion rate, adding DDGS to an animal's diet may negate the need for supplemental phosphorus

(Tjardes and Wright, 2002). Since DDGS go through a drying process, overheating may occur and potentially cause a chemical reaction detrimental to DDGS feeding quality. In such cases, some of the carbohydrates and protein in DDGS may become chemically bound, thus making the product indigestible to the animal. Consequently, a lighter colored DDGS may generally be preferable to a darker one that is associated with heat damage.[8]

DDGS can also contain more sulfur than corn, thereby adding significant amounts of sulfur to the diet (Berger and Good, 2007). Sulfuric acid may be used during fermentation of the ethanol feedstock mash for pH adjustment, but that process can increase the sulfur content of the distillers' grains. If cattle consume more than 0.4 percent sulfur (dry matter) from feed and water, they may contract polioencephalomalacia.9, 10, 11 Some feeders add thiamine to reduce the risk of this disorder, but the proper inclusion level of thiamine and the likelihood of it completely eliminating the disorder is not certain. In addition, excessive sulfur interferes with an animal's ability to absorb copper and its metabolic rate. Thus, in geographic regions with high levels of sulfur in forages and water, feeders may need to reduce the levels of DDGS added to diets.

As mentioned previously, phosphorus levels in DDGS (0.89 percent) are greater than those in corn (0.25 percent) (see table 1), so adding DDGS to an animal's diet may negate the need for phosphorus supplements, which are costly. Phosphorus concentrations may determine inclusion rate in many diets where nutrient management of the waste is a problem.[12] Research is being conducted to develop methods for removing phosphorus from DDGS (Berger and Good, 2007).

The sodium content of DDGS may vary from 0.01 percent to 0.48 percent, averaging 0.11 percent (Shurson and Alghandi, 2008). In comparison, corn contains about 0.02 percent of sodium. Salt is formed as a result of pH adjustments during processing. Salt contains a large amount of sodium and, if poultry are fed sodium above required levels, the resulting increased water consumption may cause wet litter and dirty eggs. Wet litter can encourage greater bacterial growth, increasing an animal's susceptibility to intestinal infections.

Mycotoxins—toxic chemical compounds produced by certain fungi—are also a concern for livestock/poultry feeders (USDA/GIPSA, 2006). Mycotoxins also may be associated with corn ear rot diseases and may be pathogenic for animals and humans (Siegel, 2010). If present in corn, mycotoxins become concentrated in DDGS approximately three-fold during the fermentation process. In addition, mycotoxins can be produced during storage if the distillers' grains are allowed to mold (Whitlow, 2008). The U.S. Food and Drug Administration (FDA) has been responsible for establishing and monitoring acceptable levels of mycotoxins and antibiotics in feedstuffs since 1965 (FDA, 2006). The FDA encourages livestock/poultry feeders to test their feed ingredients, ensuring that mycotoxins do not exceed acceptable levels.

Table 1. Nutrient profiles of selected feedstuffs[1]

Nutrients	Distillers' dried grains with solubles (corn)	Corn (yellow grain)	Soybean meal (dehulled, solvent)	Corn gluten meal	Corn gluten feed	Hominy	Cotton-seed meal (41 percent solvent)	Wheat middlings
	Percent							
Dry matter[2]	92.00	87.00	88.00	90.00	88.00	89.00	90.00	89.00
Crude protein[3]	27.00	7.50	47.80	60.00	21.00	11.50	41.00	15.00
Ether extract[3]	9.00	3.50	1.00	2.00	2.00	6.50	2.10	3.60
Crude fiber[3]	8.50	1.90	3.00	2.50	10.00	5.00	11.30	8.50
Calcium[3]	0.14	0.01	0.20	0.02	0.20	0.05	0.16	0.15
Phosphorus[3]	0.89	0.25	0.65	0.50	0.90	0.50	1.00	0.91
Ruminant digestible protein[3]	21.10	5.80	46.60	47.40	19.30	8.00	29.50	12.20
Ruminant TDN[3]	82.00	80.00	79.00	86.00	75.00	86.00	72.00	81.00
Metabolizable energy for poultry								
Kcal/lb	1,270	1,540	1,115	1,700	795	1,390	915	950
Kcal/kg	2,800	3,390	2,458	3,740	1,750	3,060	2,010	2,090
Metabolizable energy for swine								
Kcal/lb	1,497	1,520	1,425	n/a	1,090	1,530	1,225	1,000
Kcal/kg	3,300	3,350	3,140	n/a	2,400	3,365	2,690	2,200
	Percent							
Methionine (percent)[3]	0.51	0.18	0.70	1.90	0.50	0.22	0.51	0.12
Lysine (percent)[3]	0.80	0.24	3.02	1.00	0.60	0.45	1.70	0.70
Sulfur (percent)[3]	0.30	0.19	0.43	0.50	0.16	4	0.30	0.16

Kcal/lb=Kilocalories per pound. Kcal/kg=Kilocalories per kilogram. n/a = Data not available.

[1]All data other than dry matter content are expressed in an "as fed" basis.

[2]Dry matter content of the unit of feed being examined. [3]Nutrient content of the unit of feed being examined. [4]Insignificant amount.

Source: Lundeen, p.16, 2011.

Antibiotics also may be an issue when feeding DDGS (National Grain and Feed Association, 2009). Some U.S. ethanol plants may be using antibiotics as antimicrobials during the fermentation process to control bacterial contamination. The Food and Drug Administration's (FDA) Center for Veterinary

Medicine (CVM) is concerned about the possibility of antibiotic residues in DDGS and potential animal and human health consequences (FDA, 2009a). FDA developed and made available a method for the screening, confirmation, and determination of thirteen antibiotics in DDGS (FDA, 2009b). When the distillers' grains are used as a feed or animal food ingredient, the antimicrobial is regulated by the FDA as a food additive (FDA, 2011) unless it is generally recognized as safe (GRAS) for that use (FDA, 2010). In the past, the Center for Veterinary Medicine (CVM) did not object to specific uses of antibiotics through enforcement discretion which was provided on a temporary basis. This enforcement discretion has expired and currently no antibiotic residues are allowed in distillers' grains intended for use as a feed ingredient.

Understanding DDGS Nutrient Variability Is Essential

Because DDGS's nutritional content varies, feeders are urged to use caution before adding them to their animals' diets (Tjardes and Wright, 2002). Actual nutrient analyses of the coproducts intended for use from the truck or railcar can vary widely for batches from the same plant and for batches from different plants. DDGS nutrient concentrations may vary due to changes in the nutrient content of the corn (a 1-percent difference in grain content results in approximately a 3-percent difference in DDGS content) being processed due, in part, to agronomic conditions or corn variety. Additional DDGS nutrient variation may be caused by fermentation and distillation efficiencies, drying processes and temperatures, and/or the amount of condensed distillers' solubles blended into co-products (Shurson and Alghandi, 2008).

Reducing nutrient variation in DDGS has become a higher priority for ethanol producers as margins tighten and producers count on revenue derived from coproducts (DDGS). Consequently, producers strive to provide more uniform quality DDGS. In addition to sampling the specific load of purchased DDGS, feeders can obtain an idea of the DDGS nutrient content from several different sources, such as ethanol plants (University of Minnesota) or from feed analyses (Dairy One).

In addition to issues of product variability, Mathews and McConnell (2009) discuss ethanol feed coproducts in the diets of cattle (beef and dairy) and hogs. The limitations of these coproducts, such as variable moisture content, product availability, and nutrient excesses or deficiencies, affect how they must be handled and stored, impacting costs to feed buyers.

Potential Inclusion Levels of DDGS , by Type of Livestock/Poultry

The amount of DDGS that can be included in the diet of a particular type of livestock/poultry varies by its nutrient requirements, nutrient availability, and cost of alternative diet ingredients. Nutritionists typically use energy, protein, amino acid, and mineral content in balancing livestock/poultry diets.[13] The optimal choice of commodities to supply these ingredients may change over time with the changing prices of competing feed ingredients, the age of the livestock/poultry, or whether the livestock/poultry is used for breeding or market stock.

Many studies that include DDGS in the diet of a particular type of livestock/ poultry are conducted only on the basis of meeting nutrient requirements, but some may assume the cost of alternative ingredients. In this section we provide a brief review of the literature on potential inclusion levels of DDGS in diets by type of livestock and poultry.[14] Results may vary based on whether data came from university feeding trials, an experimental setting, or from actual feeding levels by industry. Potential inclusion levels are derived based on the following discussion and will be used later in this study's estimates of potential DDGS feeding by type of livestock/poultry.

Beef Cattle

DDGS are a good source of energy and protein for beef cattle in all phases of production (U.S. Grains Council, 2007). Since most of the starch in corn is converted to ethanol during the fermentation process, the fat and fiber concentrations in DDGS are increased by a factor of three compared with that in corn. DDGS contain high amounts of neutral detergent fiber (NDF) but low amounts of lignin, making DDGS a highly digestible fiber source for cattle that reduces digestive upsets compared with corn. The availability of highly digestible fiber in DDGS also allows them to serve as a partial replacement for forages and concentrates (Schingoethe, 2006).

DDGS in beef cattle diets supports inclusion for growing calves, supplementation of grazing and high-roughage diets or low phosphorus diets for beef cows, wintering cows or developing heifers, and fed cattle. DDGS can contribute to lower feed costs, fewer sub-acute acidosis occurrences than from a low-roughage diet, and improved fiber digestion in the rumen (National Corn Growers Association, 2008).[15] Growing and finishing cattle offer the largest potential use of DDGS. Feedlot diets that use DDGS at levels lower than 15-20 percent of diet dry matter serve as a protein source for the animal; at levels higher than 20 percent, DDGS serve as an energy source (Erickson et al., 2007).

Finishing cattle have been fed as much as 40 percent DDGS of diet dry matter as an energy source with excellent growth performance (table 2) (U.S. Grains Council, 2007; Klopfenstein, 2008). This inclusion rate, however, creates an excess of protein and phosphorus and may cause waste disposal issues that impact manure management plans. Feeding DDGS does not change the quality or yield of beef carcasses and has no effect on the taste or other sensory characteristics of beef (U.S. Grains Council, 2007).

Klopfenstein et al. (2008) reports on a meta-analysis where various levels of wet distillers' grains were fed to feedlot cattle. Results indicate that wet distillers' grains with solubles produced higher average daily gains and higher feed-to-gain values compared with cattle fed corn-based diets without DDGS. For example, the feeding value of wet distillers' grains with solubles at a 20-percent inclusion level was 142 percent with a decline to 131 percent at the 40-percent inclusion level. A similar analysis of dry distillers' grains with solubles showed a similar positive response but with less feeding value for dry versus wet distillers' grains. For example, the feeding value of DDGS at a 20-percent inclusion rate was 123 percent and at the 40-percent inclusion rate it declined to 100 percent. Erickson et al. (2007) reports that the biological optimum inclusion levels for dry distillers' grains with solubles (DDGS) is 20 percent for cattle on feed; however, higher levels of DDGS inclusion also provide positive feeding values. For wet distillers' grains with solubles, he reports biological optimal inclusion levels of 30 to 40 percent.

Erickson et al. (2005) suggests supplementing protein when finishing cattle diets contain less than 20 percent DDGS, as recommended by the National Research Council (2000). However, Vander Pol et al. (2005) reported that there was no benefit to supplementing finishing cattle diets with urea when diets contained 10-20 percent DDGS.

Forage diets usually maintain beef cows and replacement heifers but may require supplemental protein, energy, and phosphorus to achieve expected maintenance and growth levels. Most forage protein is degraded in the rumen, but cattle also need bypass protein (or protein not degraded in the rumen) (Stanton, 1998). DDGS provide a good source of bypass protein. DDGS fed to cattle grazing high-forage diets increases weight gains and reduces forage consumption, thereby, providing producers with an opportunity to extend the grazing period (U.S. Grains Council, 2007). The U.S. Grains Council (2007) reports that inclusion rates of 10-30 percent yielded beneficial results for beef cows and replacement heifers (table 2). Other potential uses of DDGS include feed for nursing and growing calves that require more protein. The National Corn Growers Association (2008) recommends a 10-20 percent inclusion rate for other cattle.

Table 2. Daily DDGS inclusion rates for livestock/poultry as a percent of daily dry matter intake

		Range of selected maximum potential inclusion rates	
Type of livestock/poultry	NASS actual[1]	Low	High
	Percent of dry matter intake		
Beef			
Cows	22	10	30[2]
Replacement heifers	N/A	10	30[2]
Cattle on feed	23	20	40[3]
Other	N/A	10	30[2, 4]
Dairy			
Cows	8	10	30[5]
Replacement heifers	N/A	10	30[5]
Hogs			
Breeding swine	10	10	50[2, 6]
Market swine	10	10	30[2, 6]
Poultry Layers	N/A	10	15[2, 7]
Pullets	N/A	10	15[2, 7]
Broilers	N/A	10	15[2, 7]
Turkeys	N/A	10	15[2, 7]

NASS=National Agricultural Statistics Service. N/A=Not applicable.
[1]USDA, NASS, 2007.
[2]U.S. Grains Council, 2007.
[3]Klopfenstein et al., 2008; Erickson et al., 2007.
[4]National Corn Growers Association, 2008.
[5]Schingoethe, 2008.
[6]Stein, 2008.
[7]Bregendahl, 2008.

USDA's National Agricultural Statistics Service (NASS) survey, *Ethanol Co-Products Used for Livestock Feed*, provides a 2006 estimate of annual usage and inclusion rate of distillers' grains for cattle and hogs from Midwestern feeders (USDA/NASS, 2007).[16] This survey reported that beef cattle (cow/calf) were fed an average of 396 pounds of DDGS in 2006 at a 22-percent inclusion rate (see table 2). Cattle on feed consumed an average of 916 pounds in 2006 at a dry matter inclusion rate of 23 percent.

Dairy Cattle

DDGS provide a source of protein, fat, phosphorus, and energy for dairy cows. DDGS are a particularly good source of protein for cattle that is undegradable in the rumen (or by-pass protein). DDGS provide high amounts of neutral detergent fiber but offer low amounts of lignin, making them a highly digestible fiber source that reduce digestive upset more effectively than corn. Although they usually replace concentrate ingredients, the highly digestible fiber in DDGS also serves as a partial replacement for forages and concentrates in diets for dairy cattle although they usually replace concentrate ingredients (Shingoethe, 2008). The quality of protein in DDGS is fairly good, but lysine is the first limiting amino acid. (See Shcingoethe (2008) for further discussion of dairy cattle protein needs and amino acids.) Thus, milk production can sometimes be increased when cows are fed rations containing supplemental lysine and methionine that is protected in the rumen or when DDGS are blended with other high-lysine ingredients. Feeding DDGS to dairy cattle results in milk production as high or higher than when dairy cows are fed rations containing soybean meal as the protein source (U.S. Grains Council, 2007).

DDGS inclusion levels are not the only factor to consider when formulating the dairy cow diet. Other factors that could affect milk production and milk composition when DDGS are added to the diet include the type of forage, the ratio of forage to concentrate, the high oil content of DDGS, and the formulation of diets on an amino acid basis. In addition, the nutrient differences between DDGS and WDGS may affect the cow's ability to produce milk.

DDGS can also be used in diets of dairy calves, heifers, and dry cows. Dry cows were fed about 10 percent of dry matter and calves 28 percent. Different levels of DDGS have been fed to dairy heifers along with a blending of other feeds (Schingoethe, 2008).

Milk fat content may decrease if inadequate amounts of forage fiber are fed to dairy cattle. DDGS can be included in dairy cow diets at up to 20 percent of the ration without decreasing dry matter intake, milk production, and milk fat and protein percentage (see table 2). Inclusion of DDGS at 20-30 percent also supports milk production equal to or greater than diets with no DDGS; however, milk production from cows fed diets containing WDGS decreases when fed at more than 20 percent of the diet (U.S. Grains Council, 2007). Thus, dairy producers can feed more than the typical 5-10 percent that many have been feeding. When feeding DDGS at more than 20 percent of the ration, however, DDGS lacks a nutritional advantage because such diets may contain excess protein and phosphorus, even though milk production performance is high with inclusion levels greater than 20 percent (Schingoethe, 2008).

NASS survey results suggest that Midwestern dairy cow feeders fed an average of 1,002 pounds of DDGS during 2006 at a dry matter inclusion rate of 8 percent (USDA/NASS, 2007) (see table 2).

Swine

DDGS can be used in the gestation, lactation, nursery, growing, and finishing diets for swine (Stein, 2008). DDGS can be an economical source of energy, amino acids, and phosphorus for swine. Swine, however, cannot efficiently digest the fiber in DDGS, and the corn oil present in DDGS can potentially affect meat quality. The amount of dicalcium phosphate normally fed can be reduced when feeding DDGS to swine because distillers' grains have a greater level of digestible phosphorus than corn. According to Stein (2008), DDGS included in the diet may have a positive effect on the health of the swine. For example, the incidence and severity of proliferative ileitis, an inflammation of the lower part of the small intestine common in young pigs, can be reduced by including 10 percent DDGS in feed rations.

Wilson et al. (2003) shows that DDGS can be fed to gestating sows at an inclusion rate of up to 50 percent with no negative effects on the animals. DDGS in the diets of gestating sows did not affect lactation feed intake, litter weight gain, and reproduction cycle. Negative effects were not seen in sow gestation weight gain, pigs born alive per litter, litter birth weight, or average pig birth weight for sows fed 0-50 percent of DDGS during gestation. Stein (2008) reports of research results that feeding DDGS to lactating sows at diet inclusion rates from 15 percent to 30 percent resulted in no negative effects. Thus, DDGS can be included in diets of gestating sows at inclusion rates of up to 50 percent and in diets of lactating sows at inclusion rates of up to 30 percent if diets are formulated based on concentrations of digestible energy, amino acids, and phosphorus (Stein, 2008).

Stein (2008) reports the results of eight experiments of DDGS inclusion in nursery pigs. From the day of weaning, a DDGS inclusion rate of 7.5 percent could be included in the diet without negative effects on the animals. Other findings suggest that an inclusion rate of up to 25 percent may be included in the diet during the initial 2 weeks after weaning and an inclusion rate of up to 30 percent may be used 2-3 weeks after weaning with no negative effects on pig performance.

Stein (2008) also reports on research results from feeding DDGS to grow-finish hogs. Generally, there was no change in performance by including DDGS in the diets of grow-finish hogs but there were experiments where reduced performance was observed. Stein (2008) reports of many experiments where DDGS can be included in diets to grow-finish hogs at up to 30 percent without negatively affecting hog performance. For experiments with reduced performance, a linear reduction in pig performance was reported when hogs were fed diets including 10, 20, and 30 percent DDGS. Reduced performance in these cases may have been due to reduced feed intake as a result of reduced DDGS quality or palatability. If DDGS had low lysine digestibility, pig performance could decline because lysine limits protein formation. Also, excessive protein intake could lead to reduced performance.

If this were the case, it is impossible to determine if the performance decline was due to DDGS in the diet or increased crude protein. Including crystalline lysine or tryptophan in hog diets may reduce the negative impact of increasing crude protein (Stein, 2007).

Effects of DDGS upon pig carcass composition and quality are mixed. Stein's (2008) summary of research results found a reduced dressing percentage from grow-finish hogs fed DDGS. These results may be due to increased fiber concentrations in DDGS-containing diets leading to increased intestinal tissue weight and reduced dressing percentage. While DDGS quality or diet formulation may account for these differences, further research is required to determine why the dressing percentage was reduced for some experiments. In approximately half of the experiments, however, the dressing percentage remained the same.

Stein's (2008) summary of research results also indicates that backfat thickness, lean meat percentage, and loin depth were not affected by the inclusion of corn DDGS in hog finishing diets. Some research results did show a decrease in swine belly thickness for some but not all experiments. Other findings showed that including DDGS in diets reduced swine belly firmness and increased iodine values of carcass fat (Stein, 2008). The increased iodine values of carcass fat may be due to the large quantities of unsaturated lipids present in corn DDGS, whereby the lipids are incorporated into carcass fat without hydrogenation. Increased unsaturated fatty acids reduce the firmness of the fat and increase the iodine values.

An experiment by White et al. (2007) demonstrated that the inclusion of 1 percent of conjugated linoleic acid in DDGS-containing diets during the 10 days prior to slaughter may reduce iodine values and could be used to mitigate soft fat in DDGS-fed hogs. Additional research shows that if DDGS are removed from diets in the 3-4 weeks prior to slaughter, acceptable iodine values are reported in pigs fed DDGS during early stages of growth (Hill et al., 2008; Xu et al., 2008).

The U.S. Grains Council (2007) recommends up to 30 percent inclusion of DDGS for nursery pigs. Due to concerns of reduced belly firmness and soft pork fat at higher levels of DDGS inclusion, however, the council recommends a 20-percent inclusion level for grower-finisher and developing gilts. For sows, feeders can include up to 50 percent of DDGS to gestation diets and 20 percent to lactation diets. We assumed that all diets were formulated on a digestible amino acid and available phosphorus basis.

Stein (2008) recommends that approximately 30 percent of DDGS can be included in diets fed to lactating sows, weanling pigs, and grow-finish pigs, and 50 percent can be included in gestating sow diets, assuming average or above-average-quality DDGS are used. USDA, NASS survey results from 2006 provide annual average DDGS consumption rates (60 pounds per head) and the dry matter inclusion percentage (10 percent) for hog diets (USDA/NASS, 2007).

Poultry

Corn DDGS can contribute energy, protein, and phosphorus to poultry diets (Bregendahl, 2008).[17] DDGS inclusion in poultry diets initially was set at a low level due to high fiber, poor amino acid quality, and low energy concentration. Bregendahl (2008) reports that energy and amino acid levels in DDGS, however, are higher than indicated by the National Research Council (1994). The phosphorus bioavailability found in DDGS, an economic asset, is higher than in corn and can be used to replace some supplemental phosphorus sources in the diet. Phosphorus is the third most expensive ingredient in poultry rations. Feeding DDGS to poultry may increase sodium intake, and overall sodium intake needs to be monitored in the diets for poultry. High sodium levels cause increased water consumption, potentially causing wet liter, dirty eggs, and susceptibility to intestinal infections (Bregendahl, 2008). Xanthophyll—a carotenoid pigment found in corn—is also found in corn DDGS and has been shown to improve desired egg yolk color (more yellow or red) when fed to laying hens and to increase the yellow skin color of broilers (U.S. Grains Council, 2007).

Layers

Lumpkins et al. (2005) reported that feeding 0-15 percent corn DDGS to laying hens did not affect egg production, egg weight, feed consumption, or feed utilization. Lumpkins et al. (2005) recommended feeding laying hens DDGS at no more than 10-12 percent. Roberson et al. (2005) conducted two experiments of diets that contained 0-15 percent of DDGS and focused on the effects on egg production or yolk color. Roberson et al. (2005) found that including 15 percent of DDGS in the diet did not affect egg production but, due to variable research results, recommended less than 15 percent in the laying hen diet. Both experiments used diets formulated using total amino acids.

Shurson et al. (2003b) conducted a commercial layer feeding trial in Jalisco, Mexico, to evaluate effects on egg production and egg quality by including 10 percent or more of DDGS into the layers' diet under practical feeding conditions. Shurson concluded that including 10 percent DDGS in the layers' diet can significantly improve egg production and egg yolk color.

Since these experiments, Bregendahl (2008) reported that the layer industry in the U.S. Midwest has used diets containing 5-20 percent DDGS with an average of 9 percent. These inclusion rates are affected by economics, as many commercial diets are based on a least-cost basis where the relative prices of all competing ingredients are considered. Furthermore, Bregendahl (2008) reported that the U.S. Midwest laying-hen industry fed DDGS to pullets at the same levels as routinely fed to laying hens, or up to about 15 percent.

Broilers

Lumpkins et al. (2004) focused on feeding inclusion rates of 0, 6, 12, and 18 percent DDGS to young broiler chicks. Body weight and feed utilization were not affected at up to 12 percent DDGS, but gain and feed utilization were reduced when broilers were fed at an inclusion rate of 18 percent, most likely due to an amino acid deficiency in the starter diet. Due to the high fiber content and low amino acid digestibility of DDGS, feeding high levels (25-30 percent) of DDGS to starter broilers is not recommended. Based on this study, researchers recommended a 6-percent inclusion rate of DDGS in starter diets, but grow-finish diets could contain 12-15 percent DDGS. Lumpkin's et al. (2004) study results were confirmed by feeding trials sponsored by the U.S. Grains Council (2007) and conducted in Taiwan. These feeding trials found that growth performance can be maintained when including 10 percent DDGS in the diets of starter, grower, and finisher broiler diets. Results from the Lumpkins et al. (2004) study were obtained from an experiment based on a total amino acid basis. The reduced growth performance found at high levels of DDGS inclusion may be due to amino acid deficiencies (such as lysine or arginine) because of the low amino acid digestibility of DGGS.

Wang et al. (2007a) evaluated the use of constant or increasing levels of DDGS in diets for broilers. Diets were formulated on digestible amino acid basis. Diets containing 15 percent DDGS could be fed throughout the feeding period with no adverse effects on live performance or carcass composition. Inclusion of 30 percent DDGS in the broiler diet during the starter and grower periods, however, reduced body weight, elevated feed conversion, and generally reduced breast meat yield, compared with results found for broilers fed 15 percent DDGS or broilers fed the control diet.

In another study, Wang et al. (2007b) evaluated the effects of moderate to high levels of DDGS in broiler diets and the effects of rapid and multiple changes in the level of DDGS inclusion in the diet during the growth period. Diets ranged from 0 to 30 percent DDGS inclusion and were formulated based on digestible amino acids. Broilers fed diets containing 15 percent DDGS did not differ from the control diet in terms of live performance or carcass characteristics, whether fed on a continuous basis or whether alternated weekly between a 0-15 percent inclusion rate of DDGS. Broilers fed a continuous diet with 30 percent DDGS inclusion experienced significant reductions in body weight, feed intake, and breast meat yield. Broilers fed 0-30 percent DDGS inclusion rates alternating on a weekly basis experienced live performance at about half that of broilers fed diets with inclusion rates of 0-30 percent DDGS continuously and similar to those fed 15-percent inclusion rate on a constant basis, although breast meat yield in the latter case tended to decline. Study results reflect the effective use of diets with 15 percent DDGS inclusion rates and showed that the abrupt removal of this level of DDGS did not adversely affect broiler performance.

Turkeys

Noll (2004) fed turkey toms diets up to 12 percent DDGS during the grower-finisher period and found no difference in body weight gain and feed conversion compared with turkeys fed the control corn-soybean meal diet. Also, the diets had no negative effects on breast meat yield. Roberson (2003) reported that DDGS could be included in turkey diets at the 10-percent level without affecting body weight gain or feed conversion of the turkeys, suggesting that DDGS can successfully be included at a 10-percent level for the grow-finish diets.

The U.S. Grains Council (2007) recommended maximum dietary inclusion levels for DDGS at 10 percent for broilers and turkeys and 15 percent for layers. The council added, however, that higher levels of DDGS can be used successfully with appropriate diet formulation adjustments for energy and amino acids. It further mentioned that diet formulation with DDGS should use digestible amino acid values, especially for lysine, methionine, cystine, and threonine, and minimum acceptable levels for tryptophan and arginine due to the second limiting nature of these amino acids in DDGS protein.

Bregendahl (2008) concludes that DDGS can be fed to broilers, turkeys, and laying hens at the 15-percent inclusion level or higher, when diets are formulated on a digestible-amino-acid basis. He recommends that younger broilers should receive lower inclusion levels, but inclusion levels should be increased as the broiler matures.

The USDA, NASS survey did not report on DDGS fed to poultry.

DDGS Feeding Information

Additional sources of information on feeding DDGS to livestock/poultry can also be obtained from the following references:

- Journal of Animal Science: http://jas.fass.org/
- The Journal of Applied Poultry Research: http://japr.fass.org/
- Poultry Science: http://ps.fass.org/
- International Journal of Poultry Sciences: http://www.pjbs.org/ijps/ijps.htm
- Journal of Dairy Science: http://www.journalofdairyscience.org/
- Babcock et al. (2008): http://www.card.iastate.edu/books/ distillers_grains/
- University of Minnesota: http://www.ddgs.umn.edu/
- University of Illinois: http://ilift.traill.uiuc.edu/distillers/
- University of Nebraska: http://beef.unl.edu/byproducts.shtml
- National Corn Growers Association: http://www.ncga.com/files/pdf/ DistillersGrains9-08.pdf
- U.S. Grains Council: http://www.grains.org/ ddgs-information/217-ddgs-user-handbook
- Distillers Grain Technology Council: http://www.distillersgrains.org/ feedsource/

Furthermore, information may be found at other land grant universities not listed above.

POTENTIAL U.S. FEED CONSUMPTION OF DDGS, BY TYPE OF LIVESTOCK/POULTRY

After examining DDGS inclusion rates in livestock/poultry diets, we estimated the potential daily amount of DDGS fed (as fed basis) and multiplied it by the number of days fed and the number of head fed per crop year (a proxy for the number of head fed equals the inventory number or number slaughtered per year) to arrive at an annual estimate of potential U.S. DDGS feed consumption for a given crop year.[18]

We used the low to mid-point range for maximum DDGS inclusion rates established by type of livestock/ poultry as identified previously (see table 2).[19] The optimum set of ingredients may change over time depending on changes in the prices of competing feed ingredients, the age of the livestock/poultry, or the intended use of the livestock/poultry (i.e., breeding or market stock).

Our estimates of DDGS inclusion in the diets of livestock/poultry are approximations based on literature from livestock/poultry scientists. Consequently, our estimates may not precisely match current feeding levels because higher grain prices tend to increase use of lower priced DDGS in diets of a particular group of livestock/ poultry.

The estimates of potential U.S. DDGS feed consumption are useful in two ways.

- Estimates assist with the market share calculations of U.S. DDGS fed by livestock/poultry species which will be covered later in this report.
- Estimates of potential U.S. DDGS feed consumption provide information on whether rapidly rising levels of DDGS supply will exceed potential feed use.

We relied on a fairly uniform method to estimate potential DDGS consumption:

1) Daily intake of DDGS (on an "as fed" basis) is established for each species (see Appendix A for derivation of these estimates).[20]

2) Daily intake is then multiplied by the days fed per year to derive an annual intake.
3) The annual intake estimate is multiplied by the annual head (specified as an annual inventory number or an annual number produced) of selected beef cattle, dairy cattle, swine, and poultry taken from USDA/NASS (see tables 3 through 7 for all data sources).
4) Summing for each type of livestock/poultry provides an estimation of potential U.S. DDGS consumption levels for the respective crop year.
5) Potential consumption data are based on the assumption that DDGS prices will not be a barrier to feeding DDGS. Currently, higher grain prices may increase DDGS utilization in rations.

We estimated potential DDGS consumption by type of livestock/poultry for 5 crop years—2006/07 through 2010/11. Our DDGS estimates of U.S. potential feed use for the past 5 crop years averaged 61.8 million metric tons (mmt), with a range of 60.7-63.2 mmt.

These estimates assume a near constant daily quantity of DDGS fed and a constant number of days fed for each of the crop years analyzed (first three columns are the same or nearly so in tables 3-7) and assume a variation among crop years due to different numbers of livestock/poultry fed each crop year based on changing inventory or slaughter numbers (column 4 and subsequent calculations in columns 5 and 6 may differ in tables 3 through 7).[21]

The estimated average market share of potential DDGS consumption by type of livestock/poultry for 2006/07 through 2010/11 was 49.0 percent for beef cattle, 26.3 percent for dairy cattle, 13.4 percent for swine, and 11.3 percent for poultry (table 8).

In comparison, if we use livestock/poultry numbers from USDA (February 2011) projected for crop year 2020/21 and assume DDGS inclusion rates are similar to our current analysis, we estimate a potential 64 mmt feed consumption of DDGS with a potential set of market shares similar to our average for the past 5 crop years. For example, for 2020/21, we estimated a market share of 49 percent for beef cattle, 25 percent for dairy cattle, 14 percent for hogs, and 12 percent for poultry.

Thus, this potential feed quantity and market shares are not much different than the average estimated for the past 5 crop years.

Table 3. Potential consumption of U.S. distillers' dried grains with solubles, crop year 2006/07

Type of livestock/ poultry	Potential intake of DDGS per day (as fed)[1]	Days fed per crop year	Consumption per head and per crop year[2]	Number of livestock/ poultry fed per crop year	Potential DDGS consumption	Share
	---------------*Pounds*-----------			*1,000 head*	*1,000 metric tons*	*Percent of total*
Beef					31,178	50.1
Cows	3.55	90	319.8	32,644[3]	4,735	7.6
Replacement heifers	3.89	120	466.6	5,835[3]	1,235	2.0
Cattle on feed	8.51	365	3,105.6	14,647[3]	20,633	33.2
Other	2.80	120	336.5	29,977[3, 4]	4,576	7.4
Dairy					16,154	26.0
Cows	10.05	365	3,668.6	9,145[3]	15,218	24.5
Replacement heifers	3.98	120	477.5	4,325[3]	937	1.5
Hogs					7,781	12.5
Breeding swine	1.34	365	488.6	6,133[5]	1,359	2.2
Market swine	0.68	365	249.7	56,688[5]	6,421	10.3
Poultry					7,063	11.4
Layers	0.026	365	9.4	347,063[6]	1,479	2.4
Pullets	0.017	365	6.3	100,112[6]	287	0.5
Broilers	0.022	49	1.1	8,893,746[7]	4,284	6.9
Turkeys	0.057	147	8.4	265,373[7]	1,013	1.6
				Total	62,177	100.0

[1]See Appendix A for explanation of daily DDGS potential consumption.

[2]May not be the same animal due to turnover.

[3]Inventory as of January 1, 2007.

[4]Other cattle not in feedlots include other heifers, steers, bulls over 500 pounds, and calves under 500 pounds.

[5]Average quarterly inventory for corn crop year, September-August.

[6]Average monthly (number as of beginning of month) inventory for corn crop year, September-August.

[7]Number produced during crop year (.333 x calendar year number for 2006 + .667 x calendar year number for 2007).

Source: Cattle inventory numbers for 2007 are from *Cattle Final Estimates*, 2004-2008, http://usda.mannlib.cornell.edu/usda/nass/SB989/ sb1019.pdf; Quarterly hogs and pigs data from *Hogs and Pigs Final Estimates*, 2004-2007, http://usda.mannlib.cornell.edu/usda/nass/SB986/ sb1020.pdf; Layers and pullets data from *Chicken and Egg Annual Summary*, http://usda.mannlib.cornell.edu/ MannUsda/viewDocumentInfo.do?documentID=1509; Broilers and turkeys data from P*oultry Production and Value*, http://usda.mannlib.cornell.edu/ MannUsda/viewDocumentInfo.do?documentID=1130.

Table 4. Potential consumption of U.S. distillers' dried grains with solubles, crop year 2007/08

Type of livestock/ poultry	Potential intake of DDGS per day (as fed)[1]	Days fed per crop year	Consumpti on per head and per crop year[2]	Number of livestock/ poultry fed per crop year	Potential DDGS consumption	Share
	----------------Pounds-----------			*1,000 head*	*1,000 metric tons*	*Percent of total*
Beef					31,273	49.5
Cows	3.55	90	319.8	32,435[3]	4,705	7.4
Replacement heifers	3.88	120	466.1	5,647[3]	1,194	1.9
Cattle on feed	8.51	365	3,105.6	14,827[3]	20,886	33.1
Other	2.80	120	336.0	29,454[3, 4]	4,488	7.1
Dairy					16,362	25.9
Cows	10.05	365	3,668.6	9,257[3]	15,404	24.4
Replacement heifers	3.99	120	478.3	4,415[3]	958	1.5
Hogs					8,439	13.4
Breeding swine	1.34	365	488.6	6,193[5]	1,373	2.2
Market swine	0.70	365	254.0	61,325[5]	7,067	11.2
Poultry					7,109	11.3
Layers	0.026	365	9.4	341,733[6]	1,456	2.3
Pullets	0.017	365	6.3	102,847[6]	295	0.5
Broilers	0.022	49	1.1	8,975,134[7]	4,323	6.8
Turkeys	0.057	147	8.4	271,003[7]	1,035	1.6
				Total	63,184	100.0

[1]See Appendix A for explanation of daily DDGS potential consumption.
[2]May not be the same animal due to turnover.
[3]Inventory as of January 1, 2008.
[4]Other cattle not in feedlots include other heifers, steers, bulls over 500 pounds, and calves under 500 pounds.
[5]Average quarterly inventory for corn crop year, September-August.
[6]Average monthly (number as of beginning of month) inventory for corn crop year, September-August.
[7]Number produced during crop year (.333 x calendar year number for 2006 + .667 x calendar year number for 2007).

Source: Cattle inventory numbers for 2008 are from *Cattle Final Estimates*, 2004-2008, http://usda.mannlib.cornell.edu/usda/nass/SB989/ sb1019.pdf; Quarterly hogs and pigs data from *Hogs and Pigs Final Estimates*, 2004-2007, http://usda.mannlib.cornell.edu/usda/nass/SB986/ sb1020.pdf; Layers and pullets data from *Chicken and Egg Annual Summary*, http://usda.mannlib.cornell.edu/ MannUsda/viewDocumentInfo.do?documentID=1509; Broilers and turkeys data from P*oultry Production and Value*, http://usda.mannlib.cornell.edu/Mann Usda/viewDocumentInfo.do?documentID=1130.

Table 5. Potential consumption of U.S. distillers' dried grains with solubles, crop year 2008/09

Type of livestock/poultry	Potential intake of DDGS per day (as fed)[1]	Days fed per crop year	Consumption per head and per crop year[2]	Number of livestock/ poultry fed per crop year	Potential DDGS consumption	Share
	-------------------Pounds-------------			*1,000 head*	*1,000 metric tons*	*Percent of total*
Beef					**29,831**	**48.3**
Cows	3.55	90	319.8	31,712[3]	4,600	7.4
Replacement heifers	3.89	120	466.5	5,531[3]	1,170	1.9
Cattle on feed	8.51	365	3,105.6	13,856[3]	19,519	31.6
Other	2.81	120	337.4	29,6803, 4	4,542	7.3
Dairy					**16,487**	**26.7**
Cows	10.05	365	3,668.6	9,333[3]	15,531	25.1
Replacement heifers	3.98	120	478.0	4,410[3]	956	1.5
Hogs					**8,604**	**13.9**
Breeding swine	1.34	365	488.6	6,020[5]	1,334	2.2
Market swine	0.72	365	262.8	60,974[5]	7,269	11.8
Poultry					**6,898**	**11.2**
Layers	0.026	365	9.4	337,594[6]	1,438	2.3
Pullets	0.017	365	6.3	101,215[6]	290	0.5
Broilers	0.022	49	1.1	8,703,080[7]	4,192	6.8
Turkeys	0.057	147	8.4	255,927[7]	977	1.6
				Total	**61,819**	**100.0**

[1]See Appendix A for explanation of daily DDGS potential consumption.

[2]May not be the same animal due to turnover.

[3]Inventory as of January 1, 2009.

[4]Other cattle not in feedlots include other heifers, steers, bulls over 500 pounds, and calves under 500 pounds.

[5]Average quarterly inventory for corn crop year, September-August.

[6]Average monthly (number as of beginning of month) inventory for corn crop year, September-August.

[7]Number produced during crop year (.333 x calendar year number for 2006 + .667 x calendar year number for 2007).

Source: Cattle inventory numbers for 2009 are from *Cattle,* http://usda.mannlib.cornell.edu/usda/nass/Catt//2010s/2010/Catt-01-29-2010. pdf; Quarterly hogs and pigs data from *Quarterly Hogs and Pigs report*, http://usda.mannlib.cornell.edu/MannUsda/viewDocumentInfo. do?documentID=1086; Layers and pullets data from *Chicken and Egg Annual Summary*, http://usda.mannlib.cornell.edu/Mann Usda/view DocumentInfo.do? documentID=1509; Broilers and turkeys data from P*oultry Production and Value*, http://usda.mannlib.cornell.edu/MannUsda/viewDocument Info.do?documentID=1130.

Table 6. Potential consumption of U.S. distillers' dried grains with solubles, crop year 2009/10

Type of livestock/ poultry	Potentia l intake of DDGS per day (as fed)[1]	Days fed per crop year	Consumpti on per head and per crop year[2]	Number of livestock/ poultry fed per crop year	Potential DDGS consumption	Share
	----------------*Pounds*------------			*1,000 head*	*1,000 metric tons*	*Percent of total*
Beef					29,484	48.5
Cows	3.55	90	319.8	31,3713	4,550	7.5
Replacement heifers	3.90	120	466.1	5,4513	1,156	1.9
Cattle on feed	8.51	365	3,105.6	13,6423	19,217	31.6
Other	2.81	120	336.0	29,8063, 4	4,560	7.5
Dairy					16,099	26.5
Cows	10.05	365	3,668.6	9,0863	15,119	24.9
Replacement heifers	3.98	120	477.3	4,5263	980	1.6
Hogs					8,344	13.7
Breeding swine	1.34	365	488.6	5,8375	1,294	2.1
Market swine	0.72	365	262.8	59,1385	7,050	11.6
Poultry					6,820	11.2
Layers	0.026	365	9.4	339,2846	1,446	2.4
Pullets	0.017	365	6.3	102,9276	295	0.5
Broilers	0.022	49	1.1	8,600,2257	4,143	6.8
Turkeys	0.057	147	8.4	245,2447	937	1.5
				Total	60,748	100.0

[1]See Appendix A for explanation of daily DDGS potential consumption.

[2]May not be the same animal due to turnover.

[3]Inventory as of January 1, 2010.

[4]Other cattle not in feedlots include other heifers, steers, bulls over 500 pounds, and calves under 500 pounds.

[5]Average quarterly inventory for corn crop year, September-August.

[6]Average monthly (number as of beginning of month) inventory for corn crop year, September-August.

[7]Number produced during crop year (.333 x calendar year number for 2006 + .667 x calendar year number for 2007).

Source: Cattle inventory numbers for 2010 are from *Cattle*, http://usda.mannlib.cornell.edu/MannUsda/viewDocumentInfo.do?documentID=1017; Quarterly hogs and pigs data from *Hogs and Pigs report*, http://usda.mannlib.cornell. edu/MannUsda/view DocumentInfo.do?documentID=1086; Layers and pullets data from *Chicken and Egg Annual Summary*, http://usda.mannlib.cornell.edu/Mann Usda/viewDocumentInfo.do?documentID=1509; Broilers and turkeys data from P*oultry Production and Value*, http://usda.mannlib.cornell.edu/MannUsda/viewDocumentInfo.do? documentID=1130.

Table 7. Potential consumption of U.S. distillers' dried grains with solubles, crop year 2010/11

Type of livestock/ poultry	Potential intake of DDGS per day (as fed)[1]	Days fed per crop year	Consumption per head and per crop year[2]	Number of livestock/ poultry fed per crop year	Potential DDGS consumption	Share
	--------------------Pounds-----------			*1,000 head*	*1,000 metric tons*	*Percent of total*
Beef					29,736	48.7
Cows	3.55	90	319.8	30,865[3]	4,477	7.3
Replacement heifers	3.89	120	466.9	5,158[3]	1,092	1.8
Cattle on feed	8.51	365	3,105.6	14,023[3]	19,754	32.3
Other	2.81	120	337.5	28,830[3, 4]	4,413	7.2
Dairy					16,216	26.5
Cows	10.05	365	3,668.6	9,150[3]	15,226	24.9
Replacement heifers	3.99	120	478.7	4,557[3]	990	1.6
Hogs					8,324	13.6
Breeding swine	1.34	365	488.6	5,781[5]	1,281	2.1
Market swine	0.72	365	262.8	59,075[5]	7,043	11.5
Poultry					6,831	11.2
Layers	0.026	365	9.4	339,073[6]	1,445	2.4
Pullets	0.017	365	6.3	104,204[6]	299	0.5
Broilers	0.022	49	1.1	8,625,200[7]	4,155	6.8
Turkeys	0.057	147	8.4	244,188[7]	933	1.5
				Total	61,107	100.0

[1]See Appendix A for explanation of daily DDGS potential consumption.
[2]May not be the same animal due to turnover.
[3]Inventory as of January 1, 2011.
[4]Other cattle not in feedlots include other heifers, steers, bulls over 500 pounds, and calves under 500 pounds.
[5]Average quarterly inventory for corn crop year, September-August.
[6]Average monthly (number as of beginning of month) inventory for corn crop year, September-August.
[7]Number produced during crop year (.333 x calendar year number for 2006 + .667 x calendar year number for 2007).

Source: Cattle inventory numbers for 2011 are from *Cattle*, http://usda.mannlib.cornell.edu/MannUsda/viewDocumentInfo.do?documentID=1017; Quarterly hogs and pigs data from *Hogs and Pigs report*, http://usda.mannlib.cornell.edu/ MannUsda/view DocumentInfo.do?documentID=1086; Layers and pullets data from *Chicken and Egg Annual Summary*, http://usda.mannlib.cornell.edu/Mann Usda/view DocumentInfo.do?documentID=1509; Broilers and turkeys data from P*oultry Production and Value*, http://usda.mannlib.cornell.edu/MannUsda/viewDocumentInfo.do? document ID=1130.

Table 8. Estimates of potential annual DDGS consumption, by selected crop year and type of livestock/poultry

	Crop year											
Type of live-stock/poultry	2006/07		2007/08		2008/09		2009/10		2010/11		Average of all estimates	
	1,000 metric tons	*Percent*	*1,000 metric tons*	*Percent*	*1,000 metric tons*	*Percent*	*1,000 metric tons*	*Percent*	*1,000 metric tons*	*Percent*	*1,000 metric tons*	*Percent*
Beef	31,178	50.1	31,273	49.5	29,831	48.3	29,484	48.5	29,736	48.7	30,301	49.0
Cows	4,735	7.6	4,705	7.4	4,600	7.4	4,550	7.5	4,477	7.3	4,613	7.5
Replacement heifers	1,235	2.0	1,194	1.9	1,170	1.9	1,156	1.9	1,092	1.8	1,170	1.9
Cattle on feed	20,633	33.2	20,886	33.1	19,519	31.6	19,217	31.6	19,754	32.3	20,002	32.4
Other cattle	4,576	7.4	4,488	7.1	4,542	7.3	4,560	7.5	4,413	7.2	4,516	7.3
Dairy	16,154	26.0	16,362	25.9	16,487	26.7	16,100	26.5	16,216	26.5	16,264	26.3
Cows	15,218	24.5	15,404	24.4	15,531	25.1	15,119	24.9	15,226	24.9	15,300	24.8
Replacement heifers	937	1.5	958	1.5	956	1.5	980	1.6	990	1.6	964	1.6
Hogs	7,781	12.5	8,439	13.4	8,604	13.9	8,344	13.7	8,324	13.6	8,298	13.4
Breeding swine	1,359	2.2	1,373	2.2	1,334	2.2	1,294	2.1	1,281	2.1	1,328	2.1
Market swine	6,421	10.3	7,067	11.2	7,269	11.8	7,050	11.6	7,043	11.5	6,970	11.3
Poultry	7,063	11.4	7,109	11.3	6,898	11.2	6,820	11.2	6,831	11.2	6,944	11.2
Layers	1,479	2.4	1,456	2.3	1,438	2.3	1,446	2.4	1,445	2.4	1,453	2.4
Pullets	287	0.5	295	0.5	290	0.5	295	0.5	299	0.5	293	0.5
Broilers	4,284	6.9	4,323	6.8	4,192	6.8	4,143	6.8	4,155	6.8	4,220	6.8
Turkeys	1,013	1.6	1,035	1.6	977	1.6	937	1.5	933	1.5	979	1.6
Total	62,177	100.0	63,184	100.0	61,819	100.0	60,748	100.0	61,107	100.0	61,807	100.0

Note: Totals may not add due to rounding.

Sources: USDA, Economic Research Service calculations based on data from tables 3, 4, 5, 6, and 7.

Table 9. Selected studies from the literature that estimate potential annual DDGS consumption, by type of livestock/poultry

	Dhuyvetter et al.		Berger and Good		Dooley		Fox		Average of all estimates	
	Calendar year									
Livestock class	2000-04 average		2007		2008		2008			
	1,000 metric tons	*Percent*	*1,000 metric tons*	*Percent*	*1,000 metric tons*	*Percent*	*1,000 metric tons*	*Percent*	*1,000 metric tons*	*Percent*
Beef cattle	29,446	63.1	38,709	70.0	34,524	69.1	20,774	59.1	30,863	65.4
Beef cows	9,803	21.0	10,859	19.7	5,703	11.4	4,807	13.7	7,793	16.4
Cattle on feed	12,261	26.3	15,450	28.0	16,591	33.2	12,761	36.3	14,266	30.9
Other cattle	7,381	15.8	12,400	22.4	12,230	24.5	3,206	9.1	8,804	18.0
Dairy cattle	6,276	13.5	6,779	12.3	5,347	10.7	7,693	21.9	6,524	14.6
Dairy cows	6,276	13.5	6,779	12.3	5,347	10.7	7,693	21.9	6,524	14.6
Swine	3,663	7.9	3,824	6.9	3,842	7.7	3,677	10.5	3,752	8.2
Breeding swine	1,037	2.2	1,031	1.9	973	1.9	642	1.8	921	2.0
Market swine	2,626	5.6	2,793	5.1	2,869	5.7	3,035	8.6	2,831	6.3
Poultry	7,245	15.5	5,950	10.8	6,215	12.4	3,014	8.6	5,606	11.8
Broilers	4,486	9.6	4,263	7.7	3,709	7.4	1,809	5.1	3,567	7.5
Layers	1,818	3.9	1,686	3.1	1,644	3.3	683	1.9	1,458	3.0
Pullets	161	0.3			104	0.2			133	0.1
Turkeys	780	1.7			757	1.5	522	1.5	686	1.2
Total	46,630	100.0	55,261	100.0	49,929	100.0	35,158	100.0	46,744	100.0

Note: Totals may not add due to rounding.

Sources: Dhuyvetter et al., 2005; Berger and Good, 2007; Dooley, 2008; Fox, 2008.

Our estimates of U.S. DDGS potential consumption exceeded similar estimates found in the literature. For example, some recent estimates averaged 46.7 mmt, with a range of 35.2-55.3 mmt (table 9) (Dhuyvetter et al., 2005; Berger and Good, 2007; Dooley, 2008; Fox, 2008). These potential U.S. DDGS consumption estimates from the literature were for calendar years 2007 or 2008, or an average of earlier years, 2000-2004. Our analysis focused on the 5 different crop years—2006/07 through 2010/11—that represent a period of accelerating DDGS production. In addition to the effects of using different study periods, several other items account for differences between our study estimates and those found in the literature. First, the average potential consumption for dairy cattle and hogs from these selected studies was about half the estimated potential consumption for dairy cattle and hogs in our study, primarily due to a much lower DDGS inclusion rate for dairy cattle and hogs. Also, Dooley (2008) and Fox (2008) applied a DDGS usage adoption rate of less than 100 percent by type of livestock/ poultry, which automatically would lower final estimates of potential DDGS consumption.[22] In contrast, our study did not use adoption rates due to lack of data but instead relied on DDGS potential inclusion rates that reflected the low to mid-point range of maximum DDGS inclusion rates. Our approach also recognizes the DDGS feeding constraints as identified in USDA, NASS (2007). Furthermore, our study did not include DDGS usage in sheep, goats, fish, and horses for which DDGS feeding levels have been established, nor did the Berger and Good (2007) and Dooley (2008) studies, but the Dhuyvetter et al. (2007) and Fox (2008) studies included sheep and lambs. Our estimated average potential U.S. DDGS feed consumption was 61.8 mmt for 2006/07 through 2010/11, an amount much higher than the estimated DDGS supply of 37.4 mmt for 2010/11 or a DDGS supply of 42.5 mmt from an assumed production of 15 billion gallons of corn-based ethanol.[23] These estimates also support Hoffman and Baker's (2010) findings that projected DDGS supplies would not exceed potential U.S. DDGS feed consumption. Furthermore, in addition to providing a measure of domestic feeding potential, their estimates identify a sizable export potential.

SUBSTITUTION RATES OF DDGS FOR CORN AND SOYBEAN MEAL, BY TYPE OF LIVESTOCK/POULTRY

One of the major variables needed to compute the aggregate substitution rate of DDGS for corn and soybean meal is the substitution rate by type of livestock/poultry. Substitution rates of DDGS for corn and soybean meal vary by type of livestock/poultry. This substitution calculation takes into consideration the nutrient requirements of each type of livestock/poultry. The estimates we used were taken from the literature.

Two alternative sets of estimates are provided. The first set of estimates (alternative #1) was developed based on conservative substitution assumptions. The second set (alternative #2) is considered the substitution potential for current conditions.

For example, these substitution rates reflect improved beef cattle and dairy cattle performance from feeding DDGS, which led to faster weight gain for beef cattle and increased milk production for dairy cattle. Substitution rates may change in the future based on new research from the animal science community. Based on our findings, we formed a range of substitution rates (table 10) and later demonstrate the effect that such a range has on the aggregate substitution rate of DDGS for corn and soybean meal.

Table 10. DDGS substitution rates for corn and soybean meal, by type of livestock/poultry

One pound of DDGS substitutes for how many pounds of corn and soybean meal?

	Type of livestock/poultry			
Alternative substitution rates	Beef cattle	Dairy cattle	Swine	Poultry[1]
Alternative #1			*Pounds*	
Corn	1.00	0.45	0.89	0.51
Soybean meal	0.00	0.55	0.10	0.50
Total	1.00	1.00	0.99	1.01
Alternative #2				
Corn	1.20	0.73	0.70	0.61
Soybean meal	0.00	0.63	0.30	0.44
Total	1.20	1.36	1.00	1.05

[1]Weighted average substitution rates based on layers, broilers, and turkeys. Weights based on inventory numbers or slaughter numbers for 2006/07.

Source: Alternative #1: Beef cattle data from Vander Pol et al., 2006; Trenkle, 2003. Dairy cattle data from Anderson et al., 2006. Swine data from Shurson et al., 2002 and 2003. Poultry data from Lumpkins et al., 2004 and 2005; Roberson, 2003. Alternative #2: Beef and dairy cattle data from Arora et al., 2008. Swine and poultry data from Shurson, March and August 2009.

Alternative #1 Substitution Rates—Based on the substitution rates found in table 10, 1 pound of DDGS could replace mostly corn and soybean meal and some other items, depending on the type of livestock/poultry (table 10). For example, DDGS replaced some dicalcium phosphate but also required some additional limestone in swine diets. The substitution rate of DDGS for corn or soybean meal, however, differs by each type of livestock/poultry. Therefore, to derive an aggregate substitution rate of DDGS for corn and soybean meal, substitution rates for each major type of livestock/poultry must be examined.

Beef Cattle–One pound of DDGS or WDGS can replace 1 pound of corn for 10-40 percent of diet dry matter intake and generally zero to minimal amounts of soybean meal or urea (Vander Pol et al., 2006; Trenkle, 2003).

Dairy Cattle—One pound of DDGS or WDGS can substitute for 0.45 pounds of corn and 0.55 pounds of soybean meal for up to 20 percent of the dry matter intake for dairy cattle diets (Anderson et al., 2006).[24]

Swine—One pound of DDGS can generally substitute for 0.89 pounds of corn and 0.10 pounds of soybean meal for up to 20 percent of the swine diet (nursery, grow-finish, developing gilts, and lactating sows) (Shurson et al., 2002 and 2003a).

Poultry—For layers, 1 pound of DDGS can replace about 0.46 pounds of corn and 0.54 pounds of soybean meal for up to 12-15 percent of the diet (Lumpkins et al., 2005). We applied these same substitution rates to pullets. For broilers, 1 pound of DDGS can replace about 0.51 pounds of corn and 0.50 pounds of soybean meal for up to 15 percent of the diet during the grower and finisher periods (Lumpkins et al., 2004). For turkeys, 1 pound of DDGS can replace about 0.58 pounds of corn and 0.465 pounds of soybean meal for up to 10 percent of the grow-finish diet (Roberson, 2003). A weighted average (based on inventory numbers or numbers slaughtered for 2006/07) represents all poultry, and 1 pound of DDGS can replace 0.51 pounds of corn and 0.50 pounds of soybean meal.

Alternative #2 Substitution Rates—More recent estimates of the substitution rate of DDGS for corn and soybean meal reveal that 1 pound of DDGS replaces more than a pound of corn and soybean meal combined for beef and dairy cattle (see table 10). However, 1 pound of DDGS was found to continue to replace about a pound of corn and soybean meal combined for swine and poultry.

Arora et al. (2008) found that feeding distillers' grains to beef and dairy cattle leads to improved animal performance in terms of faster weight gain for beef cattle and increased milk production for dairy cattle. Earlier estimates did not account for this performance improvement. For beef cattle, researchers estimated that 1 pound of DDGS substitutes for 1.196 pounds of corn and 0.056 pounds of urea. For dairy cattle, they reported that 1 pound of DDGS substituted for 0.731 pounds of corn and 0.633 pounds of soybean meal. Both the beef and dairy cattle estimates were weighted for the feeding of both DDGS and WDGS. Arora et al. claimed that swine growth performance remains unchanged from earlier estimates, supporting substitution rates at 1 pound of DDGS for 0.89 pounds of corn and 0.10 pounds of soybean meal for up to 15 percent of the swine ration. Their data did not provide DDGS substitution rates for poultry.

More recently, Shurson (March 2009 and August 2009) provided updated substitution rates of DDGS for corn and soybean meal by type of livestock/poultry (see table 10). Shurson's substitution estimates for both beef and dairy cattle agree with Arora et al. (2008) when accounting for substantially higher energy values, undegradeable protein in the rumen, and improved performance when feeding DDGS to cattle. Shurson's findings show that 1 pound of DDGS substitutes for 1.196 pounds of corn and 0.056 pounds of urea (nonprotein nitrogen) in beef cattle diets. Similarly, 1 pound of DDGS substitutes for 0.731 pounds of corn and 0.633 pounds of soybean meal in dairy cow diets. Shurson provides a set of substitution rates for hogs and poultry that differ from earlier estimates found in the literature. Compared with cattle, the energy value and protein quality of DDGS is lower for swine and poultry, but this fact has not limited the use of DDGS in swine and poultry rations. One advantage of feeding DDGS to swine and poultry is the reduction in inorganic phosphate supplementation. In swine diets, Shurson (March 2009) calculated that 1 pound of DDGS substitutes for 0.699 pounds of corn, 0.295 pounds of soybean meal, and 0.03 pounds of inorganic phosphate, with small supplementation of synthetic amino acids and calcium necessary. For poultry, Shurson (March 2009) found that 1 pound of DDGS can replace 0.608 pounds of corn and 0.438 pounds of soybean meal for layers. For broilers, he found that 1 pound of DDGS can replace 0.612 pounds of corn and 0.44 pounds of soybean meal. For turkeys, Shurson (March 2009) found that 1 pound of DDGS can replace 0.547 pounds of corn and 0.460 pounds of soybean meal. He then computed a simple average for a composite set of substitution rates for poultry. For example, 1 pound of DDGS replaced 0.589 pounds of corn and 0.446 pounds of soybean meal and 0.02 lbs of inorganic phosphate, or up to 20 percent of the ration, with small supplementation of synthetic amino acids, fat, and calcium necessary. Instead, we have computed a weighted average (based on inventory numbers or numbers slaughtered for 2006/07) substitution rate for poultry with 1 pound of DDGS replacing 0.61 pounds of corn and 0.44 pounds of soybean meal (see table 10).

U.S. DDGS Feed Consumption (Market Share), by Type of Livestock/Poultry

The second major variable needed to compute the aggregate substitution rate of DDGS for corn and soybean meal is the market share of DDGS consumption for each type of livestock/poultry. These estimates must be computed since an official Government source does not provide these statistics.[25] Market share estimates of U.S. DDGS feed consumption by type of livestock/poultry for 2006/07 were estimated based on information contained in the 2007 USDA/NASS survey and information contained in table 11 (table 12). Market share estimates for the remaining crop years—2007/08 through 2010/11—were estimated based on the allocation of remaining potential consumption (potential minus actual consumption) by type of livestock/ poultry multiplied by the annual increase of U.S. DDGS feed use (table 13).

Beef cattle were estimated to consume about 66 percent of the DDGS fed in 2006/07 followed by dairy cattle at 24 percent, hogs at 6 percent, and poultry at 4 percent (table 12). In 2010/11, we estimated that beef cattle continued to consume more DDGS than any other type of livestock/poultry in the 5-year analysis period. While the market share for beef cattle is estimated to have declined to 56 percent over the period, that for dairy cattle, hogs, and poultry increased. Dairy cattle ranked second in estimated consumption of U.S. DDGS during the study period and saw market share of DDGS consumption rise from 24 to 27 percent. Although DDGS were first fed to dairy cattle from beverage distilleries and now corn dry-mill ethanol plants, beef cattle show greater DDGS consumption potential than dairy cattle because beef cattle far outnumber dairy cattle. Hogs ranked third with a market share of 6 percent in 2006/07, increasing to 10 percent in 2010/11, as inclusion rates rose along with research demonstrating DDGS potential in swine diets. Poultry ranked fourth with an estimated market share of 5 percent in 2006/07, increasing to 7 percent in 2010/11. Continuing research demonstrates that DDGS inclusion rates for poultry could increase from around the 5-10 percent range to a range of 10-15 percent and above.

Table 11. Supply and disappearance of corn-based distillers' grains

	Supply									Disappearance				
	Production						Imports					Exports		
	Distillers' spent grains from beverage distilleries		Distillers' spent grains from dry-mill plants producing fuel ethanol[1]		Total		Brewers' and distillers' dregs and wastes[2,3]		Total	Feed and residual		Brewers' and distillers' dregs and wastes[2]		
Marketing year	Percent of total production		Percent of total production		Percent of total supply		Percent of total supply			Percent of total disappearance		Percent of total disappearance		Total
	mmt	*Percent*	*mmt*	*Percent*	*mmt*	*Percent*	*mmt*	*Percent*	*mmt*	*mmt*	*Percent*	*mmt*	*Percent*	*mmt*
1992/93	0.8	40	1.2	60	2.0	97	0.1	3	2.1	2.1	100	0.0	0	2.1
1993/94	0.6	27	1.6	73	2.2	96	0.1	4	2.3	2.1	92	0.2	8	2.3
1994/95	0.5	35	1.0	65	1.5	94	0.1	6	1.6	0.9	55	0.7	45	1.6
1995/96	0.8	77	0.2	23	1.0	93	0.1	7	1.1	0.5	41	0.6	59	1.1
1996/97	0.8	64	0.5	36	1.3	91	0.1	9	1.4	0.8	53	0.7	47	1.4
1997/98	0.9	52	0.8	48	1.6	94	0.1	6	1.7	1.1	65	0.6	35	1.7
1998/99	0.8	47	1.0	53	1.8	93	0.1	7	1.9	1.2	62	0.7	38	1.9
1999/00	0.9	51	0.9	49	1.7	92	0.1	8	1.9	1.1	59	0.8	41	1.9
2000/01	0.9	36	1.6	64	2.5	95	0.1	5	2.6	1.8	70	0.8	30	2.6
2001/02	0.9	31	2.0	69	2.9	96	0.1	4	3.0	2.1	72	0.9	28	3.0
2002/03	0.9	17	4.3	83	5.2	98	0.1	2	5.3	4.6	86	0.8	14	5.3
2003/04	0.9	13	6.1	87	7.0	99	0.1	1	7.1	6.4	90	0.7	10	7.1
2004/05	0.9	11	7.3	89	8.2	99	0.1	1	8.3	7.4	88	1.0	12	8.3
2005/06	0.9	9	9.5	91	10.4	99	0.1	1	10.5	9.3	88	1.2	12	10.5

Table 11. (Continued)

	Supply									Disappearance				
	Production						Imports					Exports		
	Distillers' spent grains from beverage distilleries		Distillers' spent grains from dry-mill plants producing fuel ethanol[1]		Total		Brewers' and distillers' dregs and wastes[2,3]		Total	Feed and residual		Brewers' and distillers' dregs and wastes[2]		
Marketing year	Percent of total production		Percent of total production		Percent of total supply		Percent of total supply			Percent of total disappearance		Percent of total disappearance		Total
	mmt	*Percent*	*mmt*	*Percent*	*mmt*	*Percent*	*mmt*	*Percent*	*mmt*	*mmt*	*Percent*	*mmt*	*Percent*	*mmt*
2006/07	0.9	7	13.2	93	14.1	99	0.2	1	14.3	12.5	88	1.8	12	14.3
2007/08	0.9	4	20.5	96	21.4	100	0.1	0	21.5	17.6	82	3.9	18	21.5
2008/09	0.9	3	25.6	97	26.5	99	0.3	1	26.8	21.8	81	5.0	19	26.8
2009/10	0.9	3	32.5	97	33.4	99	0.4	1	33.8	25.5	75	8.3	25	33.8
2010/114	0.9	2	36.1	98	37.0	99	0.4	1	37.4	29.1	78	8.3	22	37.4

mmt=Million metric tons.

[1] Does not account for noncorn spent grains.

[2] Assumes brewers spent grains are minor.

[3] May contain noncorn brewers' and distillers' dregs and wastes.

[4] Estimates from the September 12, 2011 WASDE and Feed Grains Database.

Source: Update of Hoffman and Baker (table 1, p. 5, 2010).

The growth potential for U.S. feed consumption of DDGS is expected to follow a similar trend, with beef cattle consuming the majority of DDGS, followed by dairy cattle, swine, and poultry (table 13).[26] Some in industry or other researchers believe that consumption of DDGS by dairy cattle is nearing maximum use levels (Deutscher, 2009; Dooley, 2008). We do not find such a constraint at this time. Dooley mentions that dairy cattle, beef cattle, hogs, and poultry will be constrained by smaller operations in their inability to use truckload lots. We would argue that smaller farms could find alternate solutions. For example, DDGS can be mixed in with the concentrate mix and delivered to the farm and stored in the working storage located on the farm. For some beef cow operations, DDGS can be fed in range cubes, stored in bunker silos as is, or mixed with other materials, such as straw or corn fodder.

Estimation Methodology

We used results from a survey that reported DDGS feeding by type of livestock to estimate consumption by type of livestock for crop year 2006/07 (USDA, NASS, 2007). We multiplied the reported amounts of DDGS fed per crop year per animal by the number of head per crop year (specifi ed as an annual inventory number or an annual number produced) and by the adoption rate (table 12). Since this survey did not report DDGS usage by poultry, we computed it as the difference between the total DDGS consumed in 2006/07 less amounts consumed by beef cattle, dairy cattle, and swine in that crop year (table 12). We assumed inclusion amounts as listed in table 3 for each of the different types of poultry and the adoption rate (8 percent) was a residual fi gure required to equal the total amount fed to all types of poultry.

The resulting market share for each class of livestock/poultry for 2006/07 was then computed by dividing the amount of DDGS fed to each type of live-stock/poultry by the estimated total U.S. DDGS feed consumption for that year (table 12).

DDGS feed use and market shares were computed for the remaining crop years 2007/08 through 2010/11 (table 13, column 1). We calculated these numbers for each crop year based on annual DDGS feed use per year (table 11) and each crop year's remaining adoption potential (potential consumption (table 13, column 2) - actual DDGS feed consumption (table 13, column 3) = remaining potential consumption) for each type of livestock/poultry (see table 13, column 4).

Table 12. Estimated actual consumption of U.S. distillers' dried grains with solubles, by type of livestock/poultry, 2006/07 crop year

	Annual DDGS		Number of livestock/	2006/07	
	consumption		poultry fed per	acutal DDGS	
Type of livestock/poultry	per head[1, 2]	Adoption rate[3]	crop year	consumption	Share
	Pounds	*Percent*	*1,000 head*	*1,000 metric tons*	*Percent of total*
Beef				8,203	65.5
Cows	396	43	32,6444	2,521	20.1
Replacement heifers	396	43	5,8354	451	3.6
Cattle on feed	916	70	14,6474	4,260	34.0
Other	166	43	29,9774, 5	971	7.7
Dairy				2,960	23.6
Cows	1002	60	9,1454	2,494	19.9
Replacement heifers	396	60	4,3254	466	3.7
Hogs				804	6.4
Breeding swine	60	47	6,1336	78	0.6
Market swine	60	47	56,6886	725	5.8
Poultry				563	4.5
Layers	9.4	87	347,0638	115	0.9
Pullets	6.3	87	100,1128	22	0.2
Broilers	1.1	87	8,893,7469	346	2.8
Turkeys	8.4	87	265,3739	79	0.6
			Total	12,529	100.0

[1] Numbers for beef, dairy, and hogs taken from USDA, NASS (2007). Poultry numbers (potential) taken from table 3.

[2] May not be the same animal due to turnover.

[3] Numbers for beef, dairy, and hogs taken from USDA, NASS (2007). Poultry adoption rate was computed (see text for method).

[4] Inventory as of January 1, 2007.

[5] Other cattle not in feedlots include other heifers, steers, bulls over 500 pounds, and calves under 500 pounds.

[6] Average quarterly inventory for corn crop year, September - August.

[7] Represents an adoption rate necessary for total consumption minus amounts consumed by beef, dairy, and hogs to equal the amount consumed by poultry.

[8] Average monthly (number as of beginning of month) inventory for corn crop year, September - August.

[9] Number produced during crop year (.333 x calendar year number for 2006 + .667 x calendar year number for 2007).

Source: Cattle inventory numbers for 2007 from Cattle Final Estimates, 2004-2008, http://usda.mannlib.cornell.edu/usda/nass/SB989/sb1019. pdf; Quarterly hogs and pigs data from Hogs and Pigs Final Estimates, 2004-2007, http://usda.mannlib. cornell.edu/usda/nass/SB986/ sb1020.pdf; Layers and pullets data from Chicken and Egg Annual Summary, http://usda.mannlib.cornell.edu/MannUsda/ viewDocumentInfo. do?documentID=1509; Broilers and turkeys data from Poultry Production and Value, http://usda.mannlib.cornell.edu/MannUsda/ view DocumentInfo.do?documentID=1130.

Table 13. Estimated quantity and market share of U.S. distillers grains fed, by type of livestock/poultry, crop years 2006/07 through 2010/11[1]

Crop year by type of livestock/poultry	Column 1		Column 2	Column 3	Column 4	Column 5	Column 6
	Estimated quantity and market share of DDGS fed[2]		Estimated potential DDGS to be fed	(-) Less: estimated DDGS fed2	(=) equals remaining DDGS feed potential[3]	Percent of remaining potential to be fed	Allocation of change in DDGS fed next year[4]
	mmt	*Percent*	--------------*mmt*----------------			*Percent*	*mmt*
2006/07 (from table 12)			(from table 3)				
Beef	8.2	65.5	31.2	8.2	23.0	51.5	2.6
Dairy	2.9	23.6	16.1	2.9	13.2	29.5	1.5
Swine	0.8	6.4	7.8	0.8	5.3	11.7	0.6
Poultry	0.6	4.5	7.1	0.6	3.3	7.3	0.4
Total	12.5	100.0	62.2	12.5	44.7	100.0	5.1
2007/08			(from table 4)				
Beef	10.8	61.5	31.3	10.8	20.5	50.2	2.1
Dairy	4.4	25.0	16.4	4.4	12.0	29.4	1.2
Swine	1.4	7.9	8.4	1.4	5.3	12.9	0.5
Poultry	1.0	5.5	7.1	1.0	3.1	7.5	0.3
Total	17.6	100.0	63.2	17.6	40.8	100.0	4.2
2008/09			(from table 5)				
Beef	12.9	59.3	29.8	12.9	16.9	47.5	1.8
Dairy	5.6	25.9	16.5	5.6	10.9	30.6	1.1
Swine	1.9	8.9	8.6	1.9	5.0	14.1	0.5
Poultry	1.3	5.9	6.9	1.3	2.8	7.9	0.3
Total	21.8	100.0	61.8	21.8	35.5	100.0	3.7
2009/10			(from table 6)				
Beef	14.7	57.6	29.5	14.7	14.8	47.6	1.7
Dairy	6.8	26.6	16.1	6.8	9.3	30.0	1.1
Swine	2.5	9.6	8.3	2.5	4.4	14.1	0.5
Poultry	1.6	6.2	6.8	1.6	2.6	8.4	0.3
Total	25.5	100.0	60.7	25.5	31.1	100.0	3.6
2010/11			(from table 7)				
Beef	16.4	56.4	29.7	16.4	13.3	47.3	1.1
Dairy	7.9	27.0	16.2	7.9	8.3	29.7	0.7
Swine	3.0	10.2	8.3	3.0	4.0	14.2	0.3
Poultry	1.9	6.5	6.8	1.9	2.5	8.8	0.2
Total	29.1	100.0	61.1	29.1	28.1	100.0	2.4

Table 13. (Continued)

Crop year by type of livestock/poultry	Column 1		Column 2	Column 3	Column 4	Column 5	Column 6
	Estimated quantity and market share of DDGS fed[2]		Estimated potential DDGS to be fed	(-) Less: estimated DDGS fed2	(=) equals remaining DDGS feed potential[3]	Percent of remaining potential to be fed	Allocation of change in DDGS fed next year[4]
	mmt	*Percent*	---------------*mmt*---------------			*Percent*	*mmt*
15 billion gallon ethanol	**production**	**level[5]**					
Beef	17.5	55.7		17.5			
Dairy	8.6	27.2		8.6			
Swine	3.3	10.5		3.3			
Poultry	2.1	6.6		2.1			
Total	31.5	100.0		31.5			

mmt=Million metric tons.

[1]Some columns may not sum due to rounding.

[2]Estimates of total U.S. DDGS fed comes from table 11.

[3]Remaining potential to be fed = (estimated potential to be fed minus estimated quantity fed) for beef, dairy, swine, and poultry. Remaining potential was reduced for swine and poultry. Swine was reduced by 25 percent per year because of industry concern that excessive oil would cause soft meat. Poultry was reduced by 50 percent per year because of slow adoption reported by industry.

[4]The allocation of change in DDGS fed the next year was based on the proportion of the remaining potential to be fed for the current year. [5]If one were to assume 15 billion gallons of corn-based ethanol would be produced, corn use for this dry-mill fuel ethanol could total 5.17 billion bushels (2.7 gallons ethanol per bushel of corn and 93 percent of ethanol produced from dry-mill plants) which would produce about 41.0 mmt plus 1 mmt from beverage distilleries and .5 mmt from imports = about 42.5 mmt supply. From this supply, we subtract an estimated 11 mmt of exports and arrive at an estimated 31.5 mmt of DDGS fed under these assumptions.

Source: Calculations by USDA's Economic Research Service.

The proportion of this remaining potential by type of livestock/poultry was used to allocate (see table 13, column 5) the change in DDGS feed consumption for the next year (see table 13, column 6).[27] Thus, the sum of the current year's estimated DDGS feed consumption (see table 13, column 3) and the allocated change in total consumption for the upcoming year (see table 13, column 6) was equal to the upcoming year's DDGS feed consumption (see table 13, column 1 and 3). Adoption rates were not assumed for this calculation since potential consumption by type of livestock/poultry could be greater than or less than the estimated amount. The DDGS inclusion rate used for potential DDGS consumption (see table 13, column 2) was on the low to mid-part of a range of maximum inclusion rates. Previous market experience has shown that both DDGS inclusion levels and adoption increased when grain prices are higher.[28]

The share of remaining potential to be fed by type of livestock/poultry is shown in table 13, column 5. These numbers tend to differ somewhat by crop year but were used to allocate the change in annual feed consumption of DDGS to the different types of livestock/poultry. For example, in 2010/11, the percentage of remaining potential to be fed for beef, dairy, swine, and poultry was 48, 30, 14, and 9 percent, respectively. These shares are then used to allocate, by type of livestock/poultry, the increase in DDGS feed consumption for 2011/12.

As seen in table 13, actual U.S. feed use of DDGS by type of livestock/ poultry remains below U.S. potential feed use. We use this remaining potential feed consumption to allocate the annual increase in DDGS feed consumption by type of livestock/poultry. Others have allocated annual increases in domestic DDGS feed consumption by assuming a fixed percentage by type of livestock/poultry (see Westcott, 2008). In contrast, Dooley (2008) reduced his potential DDGS feed consumption numbers by assuming farms will require a minimum of a truckload, creating an adoption rate by type of livestock/ poultry.

In addition to our estimates of DDGS consumption (market share) by type of livestock/poultry, other estimates are available in the literature. For example, one set of estimates is available from the Renewable Fuels Association, although it is generally produced in conjunction with CHS, Inc.[29] Annual calendar year estimates can usually be obtained from annual industry outlook reports (Renewable Fuels Association, various years). For comparability with corn supply and demand estimates, we converted these estimates from a calendar year to a corn crop year (table 14).

Another set of estimates comes from Wisner (2011) and listed in table 14. Compared with the other market share estimates, we found that our calculations for beef cattle were highest in all years and those for dairy cattle were lowest for all years. For swine and poultry, our estimates tended to be in the middle of the other two estimates.

The estimates from our study and Wisner's (2011) converged for the later 2 years. Despite the various methods employed to compute these estimates, each set of market shares trends toward the past 5-year average potential market share that was computed earlier in our study—beef cattle at 49 percent, dairy cattle at 26.3 percent, swine at 13.4 percent, and poultry at 11.3 percent.

Table 14. Market share estimates of U.S. feed consumption of distillers' grains, by crop year and type of livestock/poultry

Crop year, by type of livestock/poultry	Renewable Fuels Association	Wisner (2011)	Current ERS study
2006/07		Percent	
Beef cattle	42.0	45.2	65.5
Dairy cattle	43.3	42.8	23.6
Swine	10.4	6.0	6.4
Poultry	4.3	6.0	4.5
Total	100.0	100.0	100.0
2007/08			
Beef cattle	39.3	47.6	61.5
Dairy cattle	42.0	40.4	25.0
Swine	13.0	6.0	7.9
Poultry	5.7	6.0	5.5
Total	100.0	100.0	100.0
2008/09			
Beef cattle	38.3	50.0	59.3
Dairy cattle	40.3	38.0	25.9
Swine	14.7	6.0	8.9
Poultry	6.7	6.0	5.9
Total	100.0	100.0	100.0
2009/10			
Beef cattle	40.4	52.4	57.6
Dairy cattle	39.4	35.6	26.6
Swine	11.8	6.0	9.6
Poultry	8.4	6.0	6.2
Total	100	100.0	100.0
2010/11[1]			
Beef cattle	41.4	53.4	56.4
Dairy cattle	39.4	34.1	27.0
Swine	10.1	6.9	10.2
Poultry	9.1	5.6	6.5
Total	100	100.0	100.0

[1]Renewable Fuels Association's estimates for 2010/11 are based on calendar year 2010. Source: Renewable Fuels Association outlook reports, various years; Wisner (2011); table 13.

Aggregate Substitution of Distillers' Grains for Corn and Soybean Meal

How does the substitution of DDGS for corn and soybean meal affect the total U.S. feed supply? A bushel of corn used for dry-mill ethanol production results in a feed coproduct (DDGS) that is about equal to a third of the corn's original weight. What is less understood, however, is how this coproduct substitutes for corn and soybean meal differently for each type of livestock/ poultry and also in aggregate across all types of livestock/poultry. Estimating this aggregate rate requires a summation of the weighted average of market share and substitution rate for each type of livestock/poultry. We provide a range of estimates for both sets of substitution rates—alternative #1 and alternative #2 (see table 10)—and for the three different sets of market share estimates (see table 14), since it is imperative to show how these different estimates affect aggregate substitution of DDGS for corn and soybean meal.

Using the substitution rates of DDGS for corn and soybean meal, by type of livestock/poultry and the estimated actual DDGS consumed (market share) by type of livestock/poultry, we computed the aggregate substitution rate for crop years 2006/07 through 20010/11 (tables 15-19). Our findings illustrate the effects of a range of substitution rates by type of livestock/poultry, but our study results focus on substitution rates from alternative #2. These substitution rates are deemed to be the most current and have been verified by an animal scientist (Shurson, March 2009 and September 2009).[30] We also computed the aggregate substitution rate based on a set of market share estimates from the Renewable Fuels Association (appendix tables 1-5) and Wisner (2011) (appendix tables 6-10), multiplied by our range of substitution rates by type of livestock/poultry, but later in the study focus only on our study's results from alternative #2 set of substitution rates. We then derived some generalizations from the range of these estimates (table 20).

Based on our study's estimates, we find that, on average, for the past 5 crop years (2006/07-2010/11), 1 metric ton (mt) of DDGS can substitute for about 1.22 mt of corn and soybean meal combined in the United States (table 20).[31] We also find that not only does the aggregate substitution rate change over time but also the amount of corn and soybean meal that is substituted for by DDGS. For example, 12.5 mmt of U.S. DDGS substitutes for 12.8 mmt of corn and 2.3 mmt of soybean meal in 2006/07 (see table 15).

Table 15. Total U.S. DDGS substitution for corn and soybean meal, by livestock/poultry type, crop year 2006/07

	Type of livestock/poultry				
	Beef cattle	Dairy cattle	Swine	Poultry	Total
Estimated:					
Market share (percent)	65.5	23.6	6.4	4.5	100.0
Consumption (mmt)	8.2	2.9	0.8	0.6	12.5
Substitution rate for...1					
1 pound of DDGS substitutes for how many pounds of corn:			Pounds		
Alternative #1	1.00	0.45	0.89	0.51	0.84
Alternative #2	1.20	0.73	0.70	0.61	1.03
1 pound of DDGS substitutes for how many pounds of soybean meal:					
Alternative #1	0.00	0.55	0.10	0.50	0.16
Alternative #2	0.00	0.63	0.30	0.44	0.19
12.5 mmt of DDGS substitutes for how many mmt of corn:			mmt		
Alternative #1	8.2	1.3	0.7	0.3	10.5
Alternative #2	9.8	2.1	0.6	0.4	12.8
12.5 mmt of DDGS substitutes for how many mmt of soybean meal:					
Alternative #1	0.0	1.6	0.1	0.3	2.0
Alternative #2	0.0	1.8	0.2	0.3	2.3
mmt=Million metric tons.					

[1]Converting from pounds to metric tons, the substitution rate becomes—Alternative #1: Substitution rate = 1 metric ton of DDGS substitutes for 1 metric ton of feed consisting of 0.84 metric ton of corn and 0.16 metric ton of soybean meal. Alternative #2: Substitution rate = 1 metric ton of DDGS substitutes for 1.22 metric ton of feed consisting of 1.03 metric ton of corn and 0.19 metric ton of soybean meal.

Source: Tables 10 and 13.

Thus, each metric ton of DDGS substitutes for 1.22 mt of feed consisting of 1.03 mt of corn and 0.19 mt of soybean meal. For the current crop year, 2010/11, our estimates suggest that 29.1 mmt of U.S. DDGS substitutes for 28.6 mmt of corn and 6.7 mmt of soybean meal. Thus for 2010/11, our aggregate substitution rate drops slightly to 1.21 compared with 1.22 for 2006/07. One mt of DDGS substitutes for 1.21 mt of corn and soybean meal combined, but the composition of feed (corn and soybean meal) substituted for changes to 0.98 mt of corn and 0.23 mt of soybean meal (table 19).

Table 16. Total U.S. DDGS substitution for corn and soybean meal, by livestock/poultry type, crop year 2007/08

	Type of livestock/poultry				
	Beef cattle	Dairy cattle	Swine	Poultry	Total
Estimated:					
Market share (percent)	61.5	25.0	7.9	5.5	100.0
Consumption (mmt)	10.8	4.4	1.4	1.0	17.6
Substitution rate for...1					
1 pound of DDGS substitutes for how many pounds of corn:			Pounds		
Alternative #1	1.00	0.45	0.89	0.51	0.83
Alternative #2	1.20	0.73	0.70	0.61	1.01
1 pound of DDGS substitutes for how many pounds of soybean meal:					
Alternative #1	0.00	0.55	0.10	0.50	0.17
Alternative #2	0.00	0.63	0.30	0.44	0.21
17.6 mmt of DDGS substitutes for how many mmt of corn:			mmt		
Alternative #1	10.8	2.0	1.3	0.5	14.5
Alternative #2	12.9	3.2	1.0	0.6	17.8
17.6 mmt of DDGS substitutes for how many mmt of soybean meal:					
Alternative #1	0.0	2.4	0.1	0.5	3.1
Alternative #2	0.0	2.8	0.4	0.4	3.6
mmt=Million metric tons.					

[1]Converting from pounds to metric tons, the substitution rate becomes—Alternative #1: Substitution rate = 1 metric ton of DDGS substitutes for 1 metric ton of feed consisting of 0.83 metric ton of corn and 0.17 metric ton of soybean meal. Alternative #2: Substitution rate = 1 metric ton of DDGS substitutes for 1.22 metric ton of feed consisting of 1.01 metric ton of corn and 0.21 metric ton of soybean meal.

Source: Tables 10 and 13.

As the use of U.S. DDGS increases in each crop year (2006/07-2010/11), the aggregate substitution rate of 1 mt of DDGS for combined corn and soybean meal declines slightly from 1.22 to 1.21 mt (see table 20). This slight decline in aggregate substitution rate occurs because of the shift in market shares from beef cattle to dairy cattle, hogs, and poultry (see tables 15 through 19). The composition of feed (corn and soybean meal) substituted for also changes. This change is due to the smaller share of DDGS going to beef cattle over time, but also an increasing share to dairy cattle, swine, and poultry (see tables 15-19). Beef cattle have the largest substitution rate for corn of any type of livestock/poultry, but the smallest substitution rate for soybean meal.

Table 17. Total U.S. DDGS substitution for corn and soybean meal, by livestock/poultry type, crop year 2008/09

	Type of livestock/poultry				
	Beef cattle	Dairy cattle	Swine	Poultry	Total
Estimated:					
Market share (percent)	59.3	25.9	8.9	5.9	100.0
Consumption (mmt)	12.9	5.6	1.9	1.3	21.8
Substitution rate for...1					
1 pound of DDGS substitutes for how many pounds of corn:			Pounds		
Alternative #1	1.00	0.45	0.89	0.51	0.82
Alternative #2	1.20	0.73	0.70	0.61	1.00
1 pound of DDGS substitutes for how many pounds of soybean meal:					
Alternative #1	0.00	0.55	0.10	0.50	0.18
Alternative #2	0.00	0.63	0.30	0.44	0.22
21.8 mmt of DDGS substitutes for how many mmt of corn:			mmt		
Alternative #1	12.9	2.6	1.7	0.7	17.9
Alternative #2	15.5	4.1	1.4	0.8	21.7
21.8 mmt of DDGS substitutes for how many mmt of soybean meal:					
Alternative #1	0.0	3.1	0.2	0.6	3.9
Alternative #2	0.0	3.6	0.6	0.6	4.7
mmt=Million metric tons.					

[1]Converting from pounds to metric tons, the substitution rate becomes—Alternative #1: Substitution rate = 1 metric ton of DDGS substitutes for 1 metric ton of feed consisting of 0.82 metric ton of corn and 0.18 metric ton of soybean meal. Alternative #2: Substitution rate = 1 metric ton of DDGS substitutes for 1.22 metric ton of feed consisting of 1.00 metric ton of corn and 0.22 metric ton of soybean meal.

Source: Tables 10 and 13.

Dairy cattle, swine, and poultry have a smaller substitution rate for corn than do beef cattle, but a larger substitution rate for soybean meal. This change in the aggregate substitution rate for both corn and soybean meal is expected to slow as changes in annual market shares by type of livestock/poultry stabilize.

An alternative set of aggregate substitution rates for corn and soybean meal were computed based on a different set of market share estimates provided by the Renewable Fuels Association (various outlook years) and Wisner (2011). Their market share estimates (appendix tables 1-10) suggest a larger market share of DDGS consumption for dairy, swine, and poultry, but a smaller market share for beef cattle, than do our study's estimates (tables 15-19). Consequently, their average aggregate substitution rate of 1.23 for crop years 2006/07 through 2010/11 is slightly higher than our study's average estimate of 1.22 (table 20). Their estimates suggest a smaller substitution of DDGS for corn, but a greater substitution of DDGS for soybean meal, than estimates from our analysis over the past 5 crop years. Again this is due to the larger market share of DDGS consumption for dairy, swine, and poultry, than estimates from our study. We present these alternative results to illustrate the effects of different market share estimates, but will focus on our study's estimates for the remainder of this report.

Elsewhere in the literature, other aggregate substitution rates of DDGS for corn and soybean meal were derived by various authors to estimate the impacts of ethanol production on feed use or land use. For example, Westcott (2007 and 2008) provided estimates based on 1 mt of DDGS replacing 1 mt of corn and soybean meal combined, comprised of 91.5 percent of corn and 8.5 percent of soybean meal. Shurson (March 2009 and September 2009) provided a critique for the Renewable Fuels Association of different studies that used DDGS substitution rates for corn and soybean meal relative to land-use credits associated with the low carbon fuel standard for ethanol.

Compared with our analysis that estimated an average aggregate DDGS substitution rate of 1 mt of DDGS for 1.22 mt of corn and soybean meal combined, Shurson provided a discussion of why aggregate substitution rates from other studies that ranged from 1.0 to 1.28 should be closer to his estimate of 1.23. As discussed earlier, these alternative aggregate estimates are a function of assumed DDGS substitution rates for corn and soybean meal for different types of livestock/poultry and the assumed market shares of domestic DDGS consumption by the different types of livestock/poultry.

Table 18. Total U.S. DDGS substitution for corn and soybean meal, by livestock/poultry type, crop year 2009/10

	Type of livestock/poultry				
	Beef cattle	Dairy cattle	Swine	Poultry	Total
Estimated:					
Market share (percent)	57.6	26.6	9.6	6.2	100.0
Consumption (mmt)	14.7	6.8	2.5	1.6	25.5
Substitution rate for...[1]					
1 pound of DDGS substitutes for how many pounds of corn:			*Pounds*		
Alternative #1	1.00	0.45	0.89	0.51	0.81
Alternative #2	1.20	0.73	0.70	0.61	0.99
1 pound of DDGS substitutes for how many pounds of soybean meal:					
Alternative #1	0.00	0.55	0.10	0.50	0.19
Alternative #2	0.00	0.63	0.30	0.44	0.22
25.5 mmt of DDGS substitutes for how many mmt of corn:			*mmt*		
Alternative #1	14.7	3.0	2.2	0.8	20.7
Alternative #2	17.6	4.9	1.7	0.9	25.2
25.5 mmt of DDGS substitutes for how many mmt of soybean meal:					
Alternative #1	0.0	3.7	0.2	0.8	4.8
Alternative #2	0.0	4.2	0.7	0.7	5.7
mmt=Million metric tons.					

[1]Converting from pounds to metric tons, the substitution rate becomes—Alternative #1: Substitution rate = 1 metric ton of DDGS substitutes for 1 metric ton of feed consisting of 0.81 metric ton of corn and 0.19 metric ton of soybean meal. Alternative #2: Substitution rate = 1 metric ton of DDGS substitutes for 1.21 metric ton of feed consisting of 0.99 metric ton of corn and 0.22 metric ton of soybean meal.

Source: Tables 10 and 13.

Table 19. Total U.S. DDGS substitution for corn and soybean meal, by livestock/poultry type, crop year 2010/11

	Type of livestock/poultry				
	Beef cattle	Dairy cattle	Swine	Poultry	Total
Estimated:					
Market share (percent)	56.4	27.0	10.2	6.5	100.0
Consumption (mmt)	16.4	7.9	3.0	1.9	29.1
Substitution rate for...1					
1 pound of DDGS substitutes for how many pounds of corn:			Pounds		
Alternative #1	1.00	0.45	0.89	0.51	0.81
Alternative #2	1.20	0.73	0.70	0.61	0.98
1 pound of DDGS substitutes for how many pounds of soybean meal:					
Alternative #1	0.00	0.55	0.10	0.50	0.19
Alternative #2	0.00	0.63	0.30	0.44	0.23
29.1 mmt of DDGS substitutes for how many mmt of corn:			mmt		
Alternative #1	16.4	3.5	2.6	1.0	23.5
Alternative #2	19.6	5.7	2.1	1.1	28.6
29.1 mmt of DDGS substitutes for how many mmt of soybean meal:					
Alternative #1	0.0	4.3	0.3	0.9	5.6
Alternative #2	0.0	5.0	0.9	0.8	6.7
mmt=Million metric tons.					

[1]Converting from pounds to metric tons, the substitution rate becomes—Alternative #1: Substitution rate = 1 metric ton of DDGS substitutes for 1 metric ton of feed consisting of 0.81 metric ton of corn and 0.19 metric ton of soybean meal. Alternative #2: Substitution rate = 1 metric ton of DDGS substitutes for 1.21 metric ton of feed consisting of 0.98 metric ton of corn and 0.23 metric ton of soybean meal.

Source: Tables 10 and 13.

EFFECTS OF DRY-MILL ETHANOL PRODUCTION ON U.S. FEED SUPPLY[32]

Concerns among livestock/poultry feeders and other segments of the public arose over the use of corn for ethanol because of the shift away from corn used for other purposes (feed, food, seed, other nonethanol industrial uses, and exports) (fig. 1).

While increased use of corn for ethanol production may be one reason that feed prices initially rose, its longer term impact on prices may be somewhat mitigated by the ethanol coproduct, DDGS, which can substitute for some of the corn used to produce ethanol but also for soybean meal. (See Trostle et al. (2011) for a longer explanation of feed and food commodity price increases.)

While ethanol expansion raised demand for corn, DDGS from the dry-mill production process provided some partially offsetting effects in the feed market. Consequently, the net effect in the domestic feed market of a bushel of corn being used for ethanol production is less than a bushel. For example, for crop year 2006/07, 1 mt of DDGS used in the domestic feed market substituted for an estimated 1.22 mt of corn and soybean meal combined, or the equivalent of about 37.8 percent (by weight) of the corn used in the associated ethanol production process.[33]

We estimated that 12.5 mmt of DDGS were fed to U.S. livestock and poultry in 2006/07 and these DDGS substituted for 12.8 mmt of corn (103 percent of domestic DDGS fed) and 2.3 mmt of soybean meal (19 percent of domestic DDGS fed) (see table 15). Similarly, in 2010/11, we estimated that 29.1 mmt of DDGS were fed to U.S. livestock and poultry and these DDGS substituted for 28.6 mmt of corn (98 percent of domestic DDGS fed) and 6.7 mmt of soybean meal (23 percent of domestic DDGS fed) (see table 18).[34] Note that for either crop year 2006/07 or 2010/11, the amount of feed (corn and soybean meal) substituted for by the DDGS represents, on a weight basis, the equivalent of about 37.8 percent of the corn used in the associated ethanol production process for the given crop year.[35]

The effects of DDGS substituting for corn and soybean meal can be seen in table 21. While there are other factors contributing to changes in the amount of U.S. corn and soybean meal fed annually, a clear increase can be seen in the amount of DDGS being fed while corn and soybean meal being fed is moderating or declining in recent years.[36]

Table 20. Aggregate substitution rates of distillers' dried grains with solubles for corn and soybean meal, by crop year and alternative substitution rate

Alternative aggregate substitution rate, by crop year	Quantity of DDGS fed in the United States	One metric ton of DDGS replaces how many metric tons of corn and soybean meal combined?	Quantity of DDGS fed in U.S. replaces how many tons of corn and soybean meal?				Total	
			Corn		Soybean meal			
			Quantity	Percent	Quantity	Percent		
	mmt	*mt*	*mmt*	*Percent*	*mmt*	*Percent*	*mmt*	*Percent*
Crop year 2006/07	12.5							
Renewable Fuels Association								
Substitution rates, alternative #1		1.00	9.1	73	3.4	27	12.5	100
Substitution rates, alternative #2		1.24	11.5	92	4.0	32	15.5	124
Wisner (2011)								
Substitution rates, alternative #1		1.00	9.0	73	3.4	27	12.5	100
Substitution rates, alternative #2		1.24	11.6	93	3.9	31	15.5	124
This study's estimates								
Substitution rates, alternative #1		1.00	10.5	84	2.0	16	12.5	100
Substitution rates, alternative #2		1.22	12.8	103	2.3	19	15.1	122
Crop year 2007/08	17.6							
Renewable Fuels Association								
Substitution rates, alternative #1		1.00	12.8	73	4.8	27	17.6	100
Substitution rates, alternative #2		1.23	15.9	90	5.8	33	21.7	123
Wisner (2011)								
Substitution rates, alternative #1		1.00	13.1	74	4.5	26	17.6	100
Substitution rates, alternative #2		1.24	16.6	94	5.2	30	21.8	124
This study's estimates								
Substitution rates, alternative #1		1.00	14.5	83	3.1	17	17.6	100
Substitution rates, alternative #2		1.22	17.8	101	3.6	21	21.4	122
Crop year 2008/09	21.8							

Table 20. (Continued)

Alternative aggregate substitution rate, by crop year	Quantity of DDGS fed in the United States	One metric ton of DDGS replaces how many metric tons of corn and soybean meal combined?	Quantity of DDGS fed in U.S. replaces how many tons of corn and soybean meal?					
			Corn		Soybean meal		Total	
			Quantity	Percent	Quantity	Percent		
	mmt	*mt*	*mmt*	*Percent*	*mmt*	*Percent*	*mmt*	*Percent*
Renewable Fuels Association								
Substitution rates, alternative #1		1.00	15.9	73	5.9	27	21.8	100
Substitution rates, alternative #2		1.22	19.6	89	7.1	33	26.7	122
Wisner (2011)								
Substitution rates, alternative #1		1.00	16.5	76	5.3	24	21.8	100
Substitution rates, alternative #2		1.23	20.8	95	6.2	28	27.0	123
This study's estimates								
Substitution rates, alternative #1		1.00	17.9	82	3.9	18	21.8	100
Substitution rates, alternative #2		1.22	21.7	100	4.7	22	26.4	122
Crop year 2009/10	25.5							
Renewable Fuels Association								
Substitution rates, alternative #1		1.00	18.6	73	6.9	27	25.5	100
Substitution rates, alternative #2		1.22	23.1	90	8.2	32	31.3	122
Wisner (2011)								
Substitution rates, alternative #1		1.00	19.6	77	5.9	23	25.5	100
Substitution rates, alternative #2		1.23	24.6	96	6.9	27	31.5	123
This study's estimates								
Substitution rates, alternative #1		1.00	20.7	81	4.8	19	25.5	100
Substitution rates, alternative #2		1.21	25.2	99	5.7	22	30.9	121
Crop year 2010/11 Renewable Fuels Association	29.1							
Substitution rates, alternative #1		1.00	21.2	73	7.9	27	29.1	100

Alternative aggregate substitution rate, by crop year	Quantity of DDGS fed in the United States	One metric ton of DDGS replaces how many metric tons of corn and soybean meal combined?	Quantity of DDGS fed in U.S. replaces how many tons of corn and soybean meal?					
			Corn		Soybean meal		Total	
			Quantity	Percent	Quantity	Percent		
	mmt	*mt*	*mmt*	*Percent*	*mmt*	*Percent*	*mmt*	*Percent*
Substitution rates, alternative #2		1.23	26.5	91	9.3	32	35.8	123
Wisner (2011)								
Substitution rates, alternative #1		1.00	22.6	78	6.5	22	29.1	100
Substitution rates, alternative #2		1.23	28.2	97	7.6	26	35.8	123
This study's estimates								
Substitution rates, alternative #1		1.00	23.5	81	5.6	19	29.1	100
Substitution rates, alternative #2		1.21	28.6	98	6.7	23	35.3	121
Average for crop year								
2006/07 to 2010/11	21.3							
Renewable Fuels Association								
Substitution rates, alternative #1		1.00	15.5	73	5.8	27	21.3	100
Substitution rates, alternative #2		1.23	19.3	90	6.9	32	26.2	123
Wisner (2011)								
Substitution rates, alternative #1		1.00	16.2	76	5.1	24	21.3	100
Substitution rates, alternative #2		1.23	20.4	95	6.0	28	26.3	123
This study's estimates								
Substitution rates, alternative #1		1.00	17.4	82	3.9	18	21.3	100
Substitution rates, alternative #2		1.22	21.2	100	4.6	21	25.8	122

mt=Metric tons.

mmt=Million metric tons.

Note: Totals may not sum due to rounding.

Source: USDA, Economic Research Service calculations based on tables 15-19 and appendix tables 1-10.

Table 21. U.S. processed feeds fed, by crop year

Feed[1] (1,000 mt)	1992/93	1993/94	1994/95	1995/96	1996/97	1997/98	1998/99	1999/00	2000/01	2001/02
Energy feeds										
Corn (September-August)	133,409	118,874	138,682	119,195	134,043	138,448	138,497	143,333	147,887	148,565
Sorghum (September-August)	11,614	11,178	9,577	7,495	13,106	9,270	6,652	7,230	5,652	5,843
Barley (September-August)	3,722	5,952	4,846	4,634	3,642	3,586	3,076	3,428	2,350	2,102
Oats (September-August)	3,625	3,924	3,432	2,773	2,880	3,088	3,036	2,809	2,946	2,250
Wheat (September-August)	3,908	9,585	7,440	6,155	7,681	8,798	6,379	8,943	5,999	3,500
Total energy feeds	156,279	149,512	163,976	140,252	161,352	163,191	157,641	165,744	164,833	162,259
Percent of total	82.5	80.0	80.8	78.3	80.0	77.6	75.5	76.4	75.8	75.0
Oilseed meals										
Soybean meal (October-September)[2]	22,000	22,936	24,079	24,141	24,784	26,213	27,812	27,529	28,706	30,001
Cottonseed meal (October-September)	1,045	2,393	2,965	2,685	2,824	2,682	2,298	2,638	2,590	3,030
Rapeseed (canola) meal (October-September)		1,002	952	1,172	1,102	1,552	1,470	1,565	1,519	1,247
Linseed meal (June-May)	92	99	92	117	135	168	153	171	178	112
Peanut meal (August-July)		103	175	164	128	86	85	125	100	131
Sunflower meal (October-September)		291	565	434	419	482	576	528	450	358
Total oilseed meals	23,138	26,824	28,827	28,713	29,392	31,183	32,394	32,556	33,543	34,879
Percent of total	12.2	14.3	14.2	16.0	14.6	14.8	15.5	15.0	15.4	16.1
Animal-protein feeds										
Meat and bone meal tankage (September-August)	2,166	2,204	2,456	2,394	2,369	2,253	2,439	2,069	1,925	1,750
Fishmeal and solubles (September-August)	422	649	338	255	282	260	296	249	214	255
Feed[1] (1,000 mt)	1992/93	1993/94	1994/95	1995/96	1996/97	1997/98	1998/99	1999/00	2000/01	2001/02

Milk products (October-September)	421	426	420	381	388	373	248	277	275	227
Total animal-protein feeds	3,008	3,280	3,214	3,030	3,039	2,886	2,982	2,595	2,413	2,232
Percent of total	1.6	1.8	1.6	1.7	1.5	1.4	1.4	1.2	1.1	1.0
Other byproduct feeds										
Wheat millfeeds (September-August)	80	238	129	142	177	4,199	6,365	6,627	6,647	6,257
Rice millfeeds (August-July)	580	592	659	600	560	545	592	650	627	618
Fats and oils (September-August)	982	1,053	1,051	1,113	1,124	1,280	1,409	1,263	1,195	1,228
Miscellaneous byproduct feeds (September-August)[3]	1,376	1,390	1,404	1,418	1,432	1,446	1,461	1,476	1,490	1,506
Total other byproduct feeds[4]	3,018	3,273	3,243	3,274	3,293	7,470	9,827	10,015	9,959	9,609
Percent of total	1.6	1.8	1.6	1.8	1.6	3.6	4.7	4.6	4.6	4.4
Grain protein feeds										
Corn gluten feed (September – August)	1,200	1,400	2,000	2,500	3,100	3,600	3,700	4,000	4,200	4,400
Corn gluten meal (September – August)	700	700	1,000	800	800	900	1,100	1,100	900	1,000
Distillers' grains (DDGS) (September – August)	2,100	2,100	900	500	800	1,100	1,200	1,100	1,800	2,100
Total grain protein feeds[5] (September – August)	4,000	4,200	3,900	3,800	4,700	5,600	6,000	6,200	6,900	7,500
Percent of total	2.1	2.2	1.9	2.1	2.3	2.7	2.9	2.9	3.2	3.5
Total feeds fed	189,442.3	187,088.7	203,160.1	179,069.2	201,776.4	210,329.5	208,843.1	217,110.0	217,648.8	216,479.0

Table 21. (Continued)

Feed[1] (1,000 mt)	2002/03	2003/04	2004/05	2005/06	2006/07	2007/08	2008/09	2009/10	2010/11
Energy feeds									
Corn (September-August)	140,934	146,850	155,838	155,330	140,726	148,793	131,625	130,174	127,006
Sorghum (September-August)	4,328	4,623	4,859	3,548	2,868	4,188	5,907	3,573	3,175
Barley (September-August)	1,999	1,723	1,428	1,041	1,360	911	990	931	1,178
Oats (September-August)	2,518	2,279	2,215	2,137	2,088	2,024	1,977	1,796	1,606
Wheat (September-August)	6,700	4,121	4,850	2,720	4,601	4,133	3,351	4,054	5,194
Total energy feeds	156,479	159,596	169,190	164,776	151,643	160,049	143,851	140,528	138,159
Percent of total	73.5	73.4	73.1	71.8	68.4	68.5	65.8	64.3	62.6
Oilseed meals									
Soybean meal (October-September)[2]	29,357	28,530	30,446	30,114	31,186	30,148	27,898	27,796	27,896
Cottonseed meal (October-September)	2,441	2,527	3,133	3,044	2,766	2,349	1,639	1,592	2,213
Rapeseed (canola) meal (October-September)	1,244	1,883	1,869	2,018	2,030	2,373	2,285	1,805	2,600
Linseed meal (June-May)	161	179	187	244	249	191	117	190	190
Peanut meal (August-July)	161	111	86	106	108	105	93	83	102
Sunflower meal (October-September)	232	317	133	278	323	311	327	368	373
Total oilseed meals	33,598	33,546	35,855	35,803	36,663	35,476	32,359	31,834	33,274
Percent of total	15.8	15.4	15.5	15.6	16.5	15.2	14.8	14.6	15.1
Animal-protein feeds									
Meat and bone meal tankage (September-August)	1,721	1,906	1,984	2,062	2,154	2,147	2,088	2,049	2,010
Fishmeal and solubles (September-August)	234	186	172	187	194	178	165	157	180
Milk products (October-September)	393	339	243	225	285	325	324	227	225

Feed1 (1,000 mt)	2002/03	2003/04	2004/05	2005/06	2006/07	2007/08	2008/09	2009/10	2010/11
Total animal-protein feeds	2,349	2,431	2,399	2,475	2,634	2,651	2,577	2,433	2,415
Percent of total	1.1	1.1	1.0	1.1	1.2	1.1	1.2	1.1	1.1
Other byproduct feeds									
Wheat millfeeds (September-August)	6,287	6,155	6,131	6,203	6,218	6,119	5,928	5,795	5,128
Rice millfeeds (August-July)	628	543	558	592	497	511	560	608	525
Fats and oils (September-August)	1,222	1,310	1,448	1,427	1,354	1,301	1,081	878	905
Miscellaneous byproduct feeds (September-August)[3]	1,521	1,536	1,552	1,567	1,583	1,588	1,592	1,597	1,600
Total other byproduct feeds[4]	9,658	9,544	9,690	9,789	9,652	9,518	9,161	8,877	8,158
Percent of total	4.5	4.4	4.2	4.3	4.4	4.1	4.2	4.1	3.7
Grain protein feeds									
Corn gluten feed (September – August)	5,300	5,300	6,000	6,300	7,800	7,800	8,000	8,300	8,400
Corn gluten meal (September – August)	1,000	800	800	900	800	700	1,000	1,100	1,100
Distillers' grains (DDGS) (September – August)	4,600	6,400	7,400	9,300	12,500	17,600	21,800	25,500	29,100
Total grain protein feeds[5] (September – August)	10,900	12,500	14,200	16,500	21,100	26,100	30,800	34,900	38,600
Percent of total	5.1	5.7	6.1	7.2	9.5	11.2	14.1	16.0	17.5
Total feeds fed	212,983.1	217,617.2	231,333.5	229,342.9	221,691.5	233,794.3	218,748.0	218,572.4	220,606.0

mt=Metric tons.

[1]Adjusted for stocks, production, foreign trade, and nonfeed uses where applicable. Latest data may be preliminary or projected.

[2]Includes use in edible soy products and shipments to U.S. territories.

[3]Includes dried beet pulp and inedible molasses due to unavailability of production data.

[4]Excludes dried beet pulp, molasses beet pulp, and inedible molasses due to unavailability of production data.

[5]Excludes brewers' dried grains.

Source: Data for feeds fed from USDA, Economic Research Service, Feed Grains Database, Yearbook Tables, (http://www.ers.usda.gov/Data/FeedGrains/Table.asp?t=29); data for grain protein feeds from Hoffman and Baker, (2010, tables 1, 5, and 6) updated as of September 12, 2011 WASDE.

Overall, in the past several years, ethanol coproduct feeds (DDGS, corn gluten feed, and corn gluten meal) have gained market share in the market for U. S. feedstuffs fed, reaching an estimated 17.5 percent in 2010/11. Most of this increased market share can be attributed to the significant increase in domestic feed consumption of DDGS in recent years. As of 2010/11, DDGS replaced soybean meal as the number two feedstuff fed, and is second only to corn.

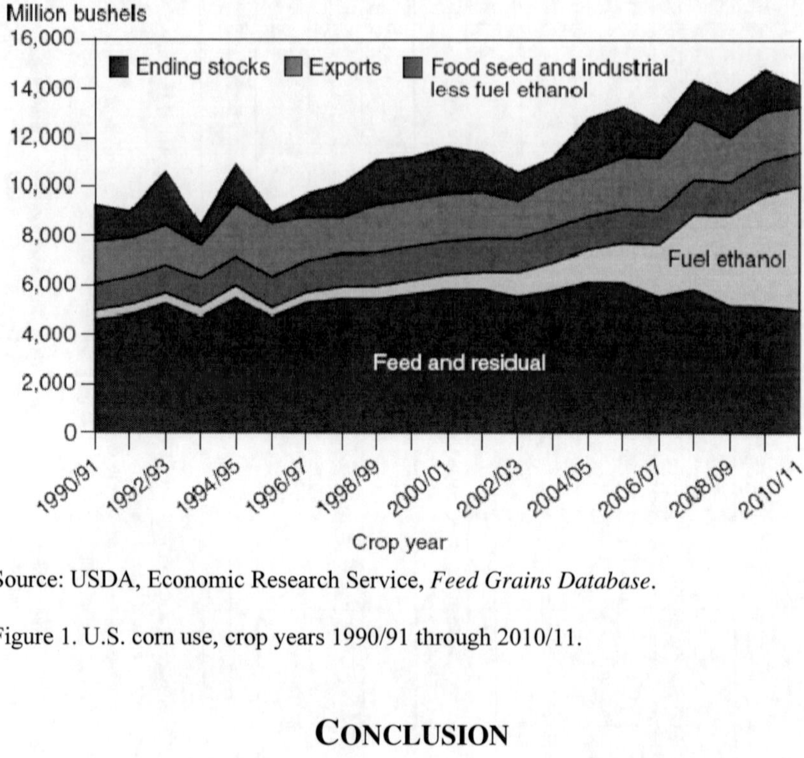

Source: USDA, Economic Research Service, *Feed Grains Database*.

Figure 1. U.S. corn use, crop years 1990/91 through 2010/11.

Conclusion

This report addressed the substitution effect of distillers' grains for corn and soybean meal on the U.S feed complex. The amount of corn and soybean meal that can be substituted in livestock/poultry diets by DDGS depends upon the actual amount (market share) of DDGS fed by type of livestock/ poultry and the substitution rates of DDGS for corn and soybean meal by type of livestock/poultry ration. We found that, on average, for the past 5 crop years (2006/07-2010/11), 1 mt of distillers' grains substitutes for about 1.22 mt of corn and soybean meal combined in the United States.

As the market shares of U.S. distillers' grains consumed by the different types of livestock/poultry change, so do the aggregate substitution rates for corn and soybean meal. For example, in 2006/07, we estimated that 1 mt of DDGS substituted for 1.22 mt of corn and soybean meal feed. Of the estimated 12.5 mmt of U.S. DDGS fed in 2006/07, DDGS substituted for 12.8 mmt of corn (103 percent of the DDGS fed) and 2.3 mmt of soybean meal (19 percent of the DDGS fed). For 2010/11, we estimate an aggregate substitution rate of 1.21 mt, but the portion of substituted corn declines and the portion of soybean meal increases. For example, we estimate that 29.1 mmt of DDGS will be fed during 2010/11 and that it will substitute for 28.6 mmt of corn (98 percent of the DDGS fed) and 6.7 mmt of soybean meal (23 percent of the DDGS fed).

Thus, aggregate substitution rates declined slightly between 2006/07 and 2010/11 but the portion of corn declined and soybean meal increased. Substitution rates changed because, as greater amounts of DDGS were fed, a declining share was estimated to be consumed by beef cattle, so relatively less corn was replaced in the U.S. feed market than was replaced by soybean meal. Furthermore, our findings show that as the market share for beef cattle declined, market shares for dairy cattle, swine, and poultry increased. Beef cattle's DDGS substitution rate for corn is higher than any other type of livestock/poultry but is the lowest for soybean meal. Changes in aggregate substitution rates are expected to slow as annual market shares by type of livestock/poultry stabilize.

Corn and soybean meal quantities fed in the United States have moderated or declined in recent years, due partly to the substitution of DDGS or other ethanol coproducts (corn gluten feed or corn gluten meal) for corn and/or soybean meal. As of 2010/11, DDGS replaced soybean meal as the number two feedstuff fed, and is second only to corn. While ethanol expansion raised demand for corn, DDGS from the dry-mill production process partially offsets the impact on the feed market. Consequently, the net effect in the domestic feed market of a bushel of corn being used for ethanol production is less than a bushel. For example, the amount of feed (corn and soybean meal) replaced by the DDGS represents about 38 percent (weight basis) of the corn used in the associated ethanol production process for a given crop year.

Future industry surveys could provide additional information on DDGS substitution for corn and soybean meal, including information on the market share of DDGS consumed by type of livestock/poultry and the substitution rates of DDGS for corn and soybean meal by type of livestock/poultry. Furthermore, if information were available for each individual ethanol coproduct—DDG, DWG, DDGS, DWGS, CDS, corn gluten feed (CGF), wet corn gluten feed (WCGF), and corn gluten meal (CGM)—estimating the effects of ethanol coproducts on the U.S. feed complex could be more precise.

REFERENCES

Anderson, J.L., D.J. Schingoethe, K.F. Kalscheur, and A.R. Hippen, "Evaluation of Dried and Wet Distillers Grains Included at Two Concentrations in the Diets of Lactating Dairy Cows," *Journal of Dairy Science* 89(8): pp. 3133–42, 2006, http://jds.fass.org/cgi/content/abstract/89/8/3133.

Arora, Salil, May Wu, and Michael Wang. *Update of Distillers Grains Displacement Ratios for Corn Ethanol Life-Cycle Analysis*, Center for Transportation Research, Energy System Division, Argonne National Laboratory, September 2008, www.transportation.anl.gov/pdfs/AF/527.pdf.

Babcock, Bruce A., Dermot J. Hayes, and John D. Lawrence (eds). *Using Distillers Grains in the U.S. and International Livestock and Poultry Industries*, Midwest Agribusiness Trade Research and Information Center at the Center for Agricultural and Rural Development, Iowa State University, 2008, http://www.card.iastate.edu/books/distillers_grains/.

Berger, Larry L., and Darrel L. Good. "Distillers Dried Grains Plus Solubles Utilization by Livestock and Poultry," *Corn-Based Ethanol in Illinois and the U.S.: A Report from the Department of Agricultural and Consumer Economics*, University of Illinois, pp. 97-111, November 2007, http://www.farmdoc.illinois.edu/policy/research_reports/ethanol_report/Ethanol%20Report.pdf.

Bregendahl, Kristjan. "Use of Distillers Co-products in Diets Fed to Poultry," *Using Distillers Grains in the U.S. and International Livestock and Poultry Industries*, Midwest Agribusiness Trade Research and Information Center at the Center for Agricultural and Rural Development, Iowa State University, pp. 99-133, 2008, http://www.card.iastate.edu/books/distillers_ grains/pdfs/chapter5.pdf.

Dahlke, Garland, and John D. Lawrence. "Ingredient Value and Cost Calculator For Livestock and Poultry Diets," *Using Distillers Grains in the U.S. and International Livestock and Poultry Industries*, Midwest Agribusiness Trade Research and Information Center at the Center for Agricultural and Rural Development, Iowa State University, pp. 179-98, 2008, http://www.card.iastate.edu/books/distillers_grains/pdfs/chapter8.pdf.

Dairy One. *Forage Feed Composition.* http://www.dairyone.com/Forage/FeedComp/MainLibrary.asp

Deutscher, Hope. "IDGC: Distillers Grains Products Markets Grow," *Ethanol Producer Magazine*, August 2009, http://www.ethanolproducer.com/articles/5816/idgc-distillers-grains-products-markets-grow.

Dhuyvetter, Kevin C., Terry L. Kastens, and Michael Boland. *The U.S. Ethanol Industry: Where will it be located in the future?* University of California, Agricultural Issues Center, November 2005, http://www.agmanager.info/agribus/energy/Ethanol%20Industry(AgMRC)—11.25.05.pdf.

Dooley, Frank J. *U.S. Market Potential for Dried Distillers Grains with Solubles*, working paper # 08-12, Department of Agricultural Economics, Purdue University, December 2008.

Ensminger, M.E., J.E. Oldfield, and W.W. Heinemann. *Feeds and Nutrition*, Second Edition, The Ensminger Publishing Company: Clovis, California, 1990.

Erickson, G.E., T.J. Klopfenstein, D.C. Adams, and R. J. Rasby. *Corn Processing Co-products Manual: A Review of Current Research on Distillers Grains and Corn Gluten*, Nebraska Corn Board and the University of Nebraska-Lincoln, Institute of Agriculture and Natural Resources, Agricultural Research Division, Cooperative Extension Division, January 2005.

Erickson, G.E., Virgil R. Bremer, Terry J. Klopfenstein, Aaron Stalker, and Rick Rasby. "Feeding of Corn Milling Co-Products to Beef Cattle," *Utilization of Corn Co-Products in the Beef Industry*, Second Edition, Nebraska Corn Board and the University of Nebraska Lincoln, Institute of Agriculture and Natural Resources, Agricultural Research Division, University of Nebraska-Lincoln Extension, August 2007, http://beef.unl.edu/byprodfeeds/manual_01_01.shtml.

Fox, John A. "The Value of Distillers Dried Grains in Large International Markets," *Using Distillers Grains in the U.S. and International Livestock and Poultry Industries*, Midwest Agribusiness Trade Research and Information Center at the Center for Agricultural and Rural Development, Iowa State University, pp. 135-53, 2008, www.card.iastate.edu/books/distillers_grains/pdfs/chapter6.pdf.

Government Printing Office. *Energy Independence and Security Act of 2007*, December 19, 2007, http://www.gpo.gov/fdsys/pkg/PLAW-110publ140/html/PLAW-110publ140.htm.

Government Printing Office. *Energy Policy Act of 2005*, August 8, 2006, http://frwebgate.access.gpo.gov/cgi-bin/getdoc.cgi?dbname=109_cong_bills&docid=f:h6enr.txt.pdf.

Hill, G.M., J.E. Link, D.O.Liptrap, M.A. Giesemann, M.J. Dawes, J.A. Snedegar, N.M.Bello, and R. J. Tempelman. "Withdrawal of Distillers Dried Grains with Solubles (DDGS) Prior to Slaughter in Finishing Pigs," *Journal of Animal Science* 86(Supplement 2): p. 52 (Abstract), 2008.

Hoffman, Linwood A., and Allen Baker. *Market Issues and Prospects for U.S. Distillers' Grains Supply, Use, and Price Relationships*, FDS-10K01, U.S. Department of Agriculture, Economic Research Service, December 2010, http://www.ers.usda.gov/Publications/FDS/2010/11Nov/ FDS10K01/.

Hutjens, Mike. *Feeding Guide*, Third Edition, W.D. Hoard and Sons Company: Fort Atkinson, Wisconsin, 2008.

Klopfenstein, T.J. Personal communication, September 20, 2010.

Klopfenstein, Terry J., Galen E. Erickson, and Virgil R. Bremer. "Use of Distillers Co-Products in Diets Fed to Beef Cattle," *Using Distillers Grains in the U.S. and International Livestock and Poultry Industries*, Midwest Agribusiness Trade Research and Information Center at the Center for Agricultural and Rural Development, Iowa State University, pp. 6-55, 2008, www.card.iastate.edu/books/distillers_grains/pdfs/ chapter2. pdf.

Lumpkins, B.S., A.B. Batal, and N.M. Dale. "Evaluation of Distillers Dried Grains with Solubles as a Feed Ingredient for Broilers," *Poultry Science* 83: pp. 1891-96, 2004, http://ps.fass.org/cgi/reprint/83/11/1891.

Lumpkins, B.S., A.B. Batal, and N.M. Dale. "Use of Distillers Dried Grains Plus Solubles in Laying Hen Diets," *The Journal of Applied Poultry Research* 14: pp. 25-31, 2005, http://japr.fass.org/cgi/content/ abstract/14/1/25.

Lundeen, Tim (ed.). "Feedstuffs Ingredients Analysis Table 2011 Edition," *Feedstuffs 2011, Reference Issue and Buyers Guide*, Minnetonka, Minnesota, p. 16, September 15, 2011.

Mathews, Kenneth H. and Michael J. McConnell. *Ethanol Co-Product Use in U.S. Cattle Feeding: Lessons Learned and Considerations*, FDS-09D-01, U.S. Department of Agriculture, Economic Research Service, April 2009, http://www.ers.usda.gov/Publications/FDS/2009/04Apr/FDS09D01/.

National Corn Growers Association. *Corn Distillers Grain: Value-added Feed Source for Beef, Dairy Beef, Dairy, Poultry, Swine, and Sheep*, September 9, 2008, http://www.ncga.com/files/pdf/DistillersGrains9-08.pdf.

National Grain and Feed Association. "FDA Sampling Detects Antibiotic Residues in Ethanol Distillers Products," *NGFA Newsletter* Volume 61, Number 2, pp. 1 and 6, January 29, 2009, http://www.ngfa.org/files/misc/News1-29-09.pdf.

National Research Council. *Nutrient Requirements of Beef Cattle*, Seventh Revised Edition: Update 2000, National Academies Press: Washington, DC, 2000, http://www.nap.edu/openbook.php?isbn=0309069343.

National Research Council. *Nutrient Requirements of Dairy Cattle*, Seventh Revised Edition, National Academies Press: Washington, DC, 2001, http://www.nap.edu/openbook.php?isbn=0309069971.

National Research Council. *Nutrient Requirements of Poultry*, Ninth Revised Edition, National Academies Press: Washington, DC, 1994, http://www.nap.edu/openbook.php?isbn=0309048923.

National Research Council. *Nutrient Requirements of Swine*, Tenth Revised Edition, National Academies Press: Washington, DC, 1998, http://www.nap.edu/openbook.php?isbn=0309059933.

Noll, S.L., J. Brannon, and V. Stangeland. "Market Turkey Performance and Inclusion Level of Corn Distillers Dried Grains with Solubles," *Poultry Science* 82(Supplement 1): p. 321, 2004.

Rausch, Kent D., and Ronald L. Belyea. "Coproducts from Bioprocessing of Corn," paper presentation at annual meeting of the American Society of Agricultural Engineers, Tampa, Florida, July 17-20, 2005, http://www.ddgs.umn.edu/articles-proc-storage-quality/2005-Rausch-%20Coproducts%20from%20bioprocessing--.pdf.

Renewable Fuels Association. 2009 *Corn Quality Issues and Considerations for Ethanol Production*, November 2009, http://ethanolrfa.3cdn.net/ffe71cb4be159b0042_njm6iyb8p.pdf.

Renewable Fuels Association. *2009 Ethanol Industry Outlook: Growing Innovation*, p. 24. 2009, http://ethanolrfa.3cdn.net/9a54006afdf842b174_rlm626rio.pdf.

Renewable Fuels Association. 2010 *Ethanol Industry Outlook: Climate of Opportunity*, p. 7, 2010, http://ethanolrfa.org/page/-/objects/pdf/outlook/RFAoutlook2010_fin.pdf?nocdn=1.

Renewable Fuels Association. 2011 *Ethanol Industry Outlook; Building Bridges to a More Sustainable Future*, p. 7, 2011, http://www.ethanolrfa.org/page/-/2011%20RFA%20Ethanol%20Industry%20Outlook.pdf?nocdn=1.

Renewable Fuels Association. *Changing the Climate: Ethanol Industry Outlook 2008*, p. 14, 2008, http://www.ethanolrfa.org/page/-/objects/pdf/outlook/RFA_Outlook_2008.pdf?nocdn=1.

Renewable Fuels Association. *Ethanol Industry Outlook 2007: Building New Horizons*, p. 13, 2007, http://www.ethanolrfa.org/page/-/objects/pdf/outlook/RFA_Outlook_2007.pdf?nocdn=1.

Roberson, K.D. "Use of Dried Distillers' Grains with Solubles in Growing-finishing Diets of Turkey Hens," *International Journal of Poultry Science* 2 (6): pp. 389-93, 2003, http://scialert.net/qredirect.php?doi=ijps.2003.389.393&linkid=pdf.

Roberson, K.D., J.L. Kalbfleisch, W. Pan, and R.A. Charbeneau. "Effect of Corn Distillers' Dried Grains with Solubles at Various Levels on Performance of Laying Hens and Egg Yolk Color," *International Journal of Poultry Science* 2: pp. 389-93, 2005.

Schingoethe, David J. "Distillers Grains for Dairy Cattle," presentation at Iowa Regional Distillers Grains Workshops, Calmar, Waverly, and Cherokee, Iowa, February 2004. http://www.ddgs.umn.edu/articles-dairy/2001-Schingoethe-%20IA%20DDGS%20workshops.pdf.

Schingoethe, David J. "Use of Distillers Co-products in Diets Fed to Dairy Cattle," *Using Distillers Grains in the U.S. and International Livestock and Poultry Industries*, Midwest Agribusiness Trade Research and Information Center at the Center for Agricultural and Rural Development, Iowa State University, pp. 57-78, 2008, http://www.card.iastate.edu/books/ distillers_grains/pdfs/chapter3.pdf.

Schingoethe, David J. "Utilization of DDGS by Cattle," presentation and paper at the 27th Western Nutrition Conference, Winnipeg, Manitoba, Canada, pp. 61-74, September 19-20, 2006, http://www.ddgs.umn.edu/articles-dairy/2006-Schingoethe-%20Utilization%20of%20DDGS%20by%20cattle.pdf.

Shurson, Jerry. "Analysis of Current Feeding Practices of Distiller's Grains with Solubles in Livestock and Poultry Feed Relative to Land Use Credits Associated with Determining the Low Carbon Fuels Standard for Ethanol," Renewable Fuel Association, March 25, 2009, http://www.ethanolrfa.org/page/-/objects/documents/2290/shurson.pdf?nocdn=1.

Shurson, Jerry, "A Scientific Assessment of the Role of Distiller's Grains (DGS) and Predictions of the Impact of Corn Co-Products Produced by Front-End Fractionation and Back-End Oil Extraction Technologies on Indirect Land Use Change," Renewable Fuels Association, September 10,

2009, http://www.ethanolrfa.org/page/-/rfa-association-site/studies/assessmentofdgs.pdf?nocdn=1.

Shurson, Jerry. *Impact of Distillers' Grains on Indirect Land Use*, presentation at the National Corn Grower Land Use Conference, St. Louis, MO, August 25-26, 2009, http://www.ddgs.umn.edu/ppt-swine/2009- Shurson-%20Impact%20of%20distillers%20grains%20on%20indirect%20land%20use%20change.pdf.

Shurson, Jerry, and Abdorrahman S. Alghandi. "Quality and New Technologies to Create Corn Co-Products from Ethanol Production," *Using Distillers Grains in the U.S. and International Livestock and Poultry Industries*, Midwest Agribusiness Trade Research and Information Center at the Center for Agricultural and Rural Development, Iowa State University, pp. 231-59, 2008, http://www.card.iastate.edu/books/distillers_grains/pdfs/chapter10.pdf.

Shurson, Jerry, and Mindy Spiehs. *Feeding Recommendations and Example Diets Containing Minnesota-South Dakota Produced DDGS for Swine*, Department of Animal Science, University of Minnesota, 2002, http://www.ddgs.umn.edu/feeding-swine/exampleswinediets-revised.pdf.

Shurson, Jerry, Mindy Spiehs, Jennifer Wilson, and Mark Whitney. "Value and use of "new generation" distillers' dried grains with solubles in swine diets," presented at the 19th International Alltech Conference, Lexington, KY, May 13, 2003a, http://www.ddgs.umn.edu/articles-swine/2003-Shurson-%2019th%20Internatl%20Alltech%20Conf.pdf.

Shurson, J., C. Santos, J. Aguirre, and S. Hernandez. "Effects of Feeding Babcock B300 Laying Hens Conventional Sanfandila Layer Diets Compared to Diets Containing 10 percent Norgold DDGS on Performance and Egg Quality," Minnesota Corn Research and Promotion Council and the Minnesota Department of Agriculture, 2003b.

Siegel, Victoria. "Corn Dried Distillers Grains and Mycotoxins," presentation made at the 14th Annual Distillers Grain Technology Council Symposium, Indianapolis, Indiana, May 12-13, 2010.

Staff, Charles, Executive Director, Distillers Grain Technology Council. Written communication, March 14, 2011.

Stanton, T.G. *Bypass Protein for Cattle no. 1.609*, Colorado State University, Fort Collins, Colorado, June 1998, http://docs.google.com/viewer?a=v&q=cache:g1Z8eXUzig8J:www.cheboygancoop.com/animalscience/beef/ 01609.pdf+bypass+protein+for+cattle,+no.+1.609.+colorado+ state+extension,+t.l.+stanton&hl=en&gl=us&pid=bl&srcid=ADGEESgF1g NWlusRRGeec26TVPMDEQZfwmb7XO5Q7UDpBnK-Kxbe6Op5cZEdjeVv

7CiCXYrefKTa3dZCEhReSDZ2tAJvupa1s-JzG2gPzyiJVllaDuvzqCmB wlyfxlA32IOjxnIodtpUNd&sig=AHIE tbSY_TsSfddwEN5k6qxw AcInld pzbA.

Stein, Hans H. "Distillers dried grains with solubles (DDGS) in diets fed to swine," Swine Focus #001, University of Illinois at Urbana-Champaign, Department of Animal Science, 2007, http://www.distillersgrains.org/files/feedsource/swine_brochure.pdf.

Stein, Hans H. "Use of Distillers Co-products in Diets Fed to Swine," *Using Distillers Grains in the U.S. and International Livestock and Poultry Industries*, Midwest Agribusiness Trade Research and Information Center at the Center for Agricultural and Rural Development, Iowa State University, pp. 79-97, 2008, www.card.iastate.edu/books/distillers_grains/pdfs/chapter4.pdf.

Tisch, David. *Animal Feeds, Feeding and Nutrition, and Ration Evaluation with CD-Rom*, Delmar Cengage Learning: Clifton Park, New York, 2006.

Tjardes, Kent, and Cody Wright. "Feeding Corn Distiller's Co-Products to Beef Cattle," ExEx 2036, South Dakota State University, College of Agriculture & Biological Sciences and USDA Cooperative Extension Service, 2002, http://pubstorage.sdstate.edu/AgBio_Publications/articles/ExEx2036.pdf.

Trenkle, Allen. "Evaluation of Wet and Dry Distillers Grains with Solubles for Finishing Holstein Steers," report to the Iowa Corn Promotion Board, November 2003, http://www.ddgs.umn.edu/articles-beef/2003- Trenkle-%20FINAL%20ISU12-03.pdf.

Trostle, Ronald, Daniel Marti, Stacey Rosen, and Paul Westcott. *Why Have Food Commodity Prices Risen Again?* WRS-1103, U.S. Department of Agriculture, Economic Research Service, June 2011, http://www.ers.usda.gov/Publications/WRS1103/.

University of Minnesota. *Distillers Grains By-products in Livestock and Poultry Feeds*, http://www.ddgs.umn.edu/.

University of Minnesota. *Nutrient Profiles: Archived Data (U.S.)*, http://www.ddgs.umn.edu/profiles-archived-us.htm.

U.S. Department of Agriculture, Economic Research Service. *Feed Grains Database*, http://www.ers.usda.gov/data/feedgrains/.

U.S. Department of Agriculture, Grain Inspection, Packers and Stockyards Administration. "What are Mycotoxins?" *Grain Fungal Diseases & Mycotoxin Reference*, pp. 7-12, September 2006, http://archive.gipsa.usda.gov/pubs/mycobook.pdf.

U.S. Department of Agriculture, National Agricultural Statistics Service. *Cattle.* January 29, 2010, http://usda.mannlib.cornell.edu/usda/nass/Catt//2010s/2010/Catt-01-29-2010.pdf.

U.S. Department of Agriculture, National Agricultural Statistics Service. *Cattle Final Estimates, 2004-2008*, SB-1019, March 2009, http://usda.mannlib.cornell.edu/usda/nass/SB989/sb1019.pdf.

U.S. Department of Agriculture, National Agricultural Statistics Service. *Chicken and Egg Annual Summary*, http://usda.mannlib.cornell.edu/MannUsda/viewDocumentInfo.do?documentID=1509.

U.S. Department of Agriculture, National Agricultural Statistics Service, *Ethanol Co-Products Used for Livestock Feed, 2007*, http://usda.mannlib.cornell.edu/usda/current/EthFeed/EthFeed-06-29-2007_revision.pdf.

U.S. Department of Agriculture, National Agricultural Statistics Service. *Hogs and Pigs: Final Estimates*, SB-1020, March 2009, http://usda.mannlib.cornell.edu/usda/nass/SB986/sb1020.pdf.

U.S. Department of Agriculture, National Agricultural Statistics Service. *Poultry Production and Value*, http://usda.mannlib.cornell.edu/MannUsda/ viewDocumentInfo.do?documentID=1130.

U.S. Department of Agriculture, National Agricultural Statistics Service. *Quarterly Hogs and Pigs*, http://usda.mannlib.cornell.edu/MannUsda/viewDocumentInfo.do?documentID=1086.

U.S. Department of Agriculture, Office of the Chief Economist, World Agricultural Outlook Board (USDA, WAOB). *USDA Agricultural Projections to 2020*, OCE-111, February 2011, http://www.usda.gov/oce/commodity/archive_projections/USDAAgriculturalProjections2019.pdf.

U.S. Department of Agriculture, Office of the Chief Economist, World Agricultural Outlook Board (USDA, WAOB). *World Agricultural Supply and Demand Estimates*, WASDE-498, September 12, 2011, http://usda.mannlib.cornell.edu/usda/current/wasde/wasde-09-12-2011.pdf.

U.S. Energy Information Administration. *Petroleum Supply Monthly*, October 28, 2010, http://www.eia.doe.gov/oil_gas/petroleum/data_publications/petroleum_supply_monthly/psm.html.

U.S. Food and Drug Administration. *Analysis of Antibiotics in Distillers Grains Using Liquid Chromatography and Ion Trap Tandem Mass Spectrometry,* David N. Heller and G.K. Hemakanthi de Alwis, LIB-4438, Office of Research, Center for Veterinary Medicine, 2009b, http://www.fda.gov/downloads/AnimalVeterinary/ScienceResearch/ToolsResources/UCM182280.pdf.

U.S. Food and Drug Administration. *Animal Veterinary, Animal Food and Feeds Product Regulation,* February 16, 2011, http://www.fda.gov/ Animal Veterinary/Products/AnimalFoodFeeds/ucm050223.htm.

U.S. Food and Drug Administration. *FY 2010 Nationwide Survey of Distillers Grains for Antibiotic Residues*, memorandum from Director, Office of Surveillance and Compliance, September 30, 2009a, http://www.fda.gov/ AnimalVeterinary/Products/AnimalFoodFeeds/Contaminants/ucm190907. htm.

U.S. Food and Drug Administration. "Mycotoxins in Feeds: CVM's Perspective," presentation made to Risk Management Agency by Michael Henry, Division of Animal Feeds, Center for Veterinary Medicine, Food and Drug Administration, August 23, 2006, http://www.fda.gov/Animal Veterinary/ Products/AnimalFoodFeeds/Contaminants/ucm050974.htm.

U.S. Food and Drug Administration, Animal Veterinary, Guidance, Compliance and Enforcement, Compliance and Enforcement. *Letter to Industry: Marketing of Animal Food Substances as Generally Recognized As Safe (GRAS)*, 2010, http://www.fda.gov/AnimalVeterinary/GuidanceComplianceEnforcement/ComplianceEnforcement/ucm229734.htm.

U.S. Grains Council. *DDGS User Handbook*, March 2007, http://www. grains.org/ddgs-information/217-ddgs-user-handbook.

Vander Pol, Kyle J., Galen E. Erickson, Matthew A. Greenquist, Terry J. Klopfenstein, and Thomas Robb. "Effect of Dietary Inclusion of Wet Distillers Grains on Feedlot Performance of Finishing Cattle and Energy Value Relative to Corn," *Nebraska Beef Cattle Reports*, University of Nebraska-Lincoln, Animal Science Department, pp. 51-3, 2006.

Vander Pol, Kyle J., Galen E. Erickson, and Terry J. Klopfenstein. "Degradable Intake Protein in Finishing Diets Containing Dried Distillers Grains," *Nebraska Beef Cattle Reports*, University of Nebraska-Lincoln, Animal Science Department, pp. 42-4, 2005.

Wang, Z., S. Cerrate, C. Coto, F. Yan, and P.W. Waldroup. "Use of Constant or Increasing Levels of Distillers Dried Grains with Solubles (DDGS) in Broiler Diets," *International Journal of Poultry Science* 6(7): pp. 501-7, 2007a.

_____, "Effect of Rapid and Multiple Changes in Level of Distillers Dried Grain with Solubles (DDGS) in Broiler Diets on Performance and Carcass Characteristics," *International Journal of Poultry Science* 6(7): pp. 725-31, 2007b.

Westcott, Paul. "Distillers Grains: Livestock Sector Uses and Implications for Feed Use of Corn and Soybean Meal," presented at World Commodity Outlook Conference, Washington, DC, May 2008.

Westcott, Paul. *Ethanol Expansion in the United States: How Will the Agricultural Sector Adjust?* FDS-07D-01, U.S. Department of Agriculture, Economic Research Service, May 2007, http://www.ers.usda.gov/publications/fds/2007/05may/fds07d01/fds07d01.pdf.

White, H., B. Richert, S. Radcliffe, A. Schnickel, and M. Latour. "Distillers Dried Grains Decreases Bacon Lean and Increases Fat Iodine Values (IV) and the Ratio of n6n:n3 but Conjugated Linoleic Acids Partially Recovers Fat Quality," *Journal of Animal Science* 85(Supplement 2): p. 78 (Abstract), 2007.

Whitlow, Lon. "Mycotoxins and Distillers Grains," *Know Mycotoxins*, North Carolina State University, http://www.knowmycotoxins.com/documents/WhitlowEnglish.pdf.

Wilson, J.A., M.H. Whitney, G.C. Shurson, and S.K. Baidoo. "Effects of Feeding Distillers Dried Grains with Solubles (DDGS) to Gestation and Lactation Diets on Reproductive Performance and Nutrient Balance in Sows," *Journal of Animal Science* 81(Supplement 2): pp. 47-8 (Abstract), 2003.

Wisner, Robert. *Estimated U.S. Dried Distillers Grains with Solubles (DDGS) Production and Use*, Iowa State University, University Extension, April 2011, http://www.extension.iastate.edu/agdm/crops/outlook/ dgsbalance sheet.pdf.

Xu, G., S.K. Baidoo, L.J. Johnston, J.E. Cannon, D. Bibus, and G.C. Shurson. "Effects of Dietary Corn Dried Distillers Grains with Solubles (DDGS) and DDGS Withdrawal Intervals, on Pig Growth Performance, Carcass Traits, and Fat Quality," *Journal of Animal Science* 86 (Supplement 2): p. 52 (Abstract), 2008.

APPENDIX A: ESTIMATING POTENTIAL DDGS USE BY LIVESTOCK/POULTRY INDUSTRY

We estimated the potential amounts of DDGS fed per corn crop year by computing the potential daily use of DDGS by livestock/poultry, multiplied by the number of days fed per year, multiplied by the number fed during the year, and then summed by type of livestock/poultry for the potential DDGS fed per year. Our potential daily inclusion levels of DDGS into the diets of livestock/poultry were taken from current literature (see table 2). Nutritionists typically use energy, protein, amino acid, and mineral content to balance livestock/poultry diets. The optimum set of ingredients may change over time depending on changing prices of competing feed ingredients, age of the livestock/poultry, or whether the livestock/poultry is used for breeding or market stock. Our estimates attempted to incorporate most of these variables. We assume that DDGS are priced so that they are economical to include in the diet of livestock/poultry. In most cases, we used a DDGS inclusion rate for the livestock/poultry diet that represented either the low end or mid-point range of maximum potential inclusion rates established in the literature (see table 2). We do not use the extreme maximum inclusion rate because we do not assume an adoption rate. Thus, we use a lower inclusion rate to compensate for the unknown adoption rate.

Beef Cattle

Cows—Our research was based on a 22.3 pound dry matter intake for a 1,175 pound beef cow in the last trimester of pregnancy (National Research Council, 2000). We assumed the cow was on a low quality grazing/high-roughage diet over the winter and in need of diet supplementation. A 20 percent DDGS inclusion rate was used for the cow on a high roughage diet (U.S. Grains Council, 2007). For the cow on a grazing diet, we assumed that 2 pounds per day of DDGS were fed (Klopfenstein, 2010). The daily potential amount of DDGS fed to beef cows (as fed) = a weighted average of 3.55 pounds (((22.3 pounds of dry matter per day x 20 percent DDGS inclusion rate (high-roughage diet)) + (2 pounds per day for the grazing diet)) x 110 percent to convert to an as fed basis. We assumed that these cows were fed over the winter for a period of 90 days. The cattle numbers fed were derived from a January 1st inventory number (USDA/NASS) (see tables 3 through 7).

Beef cow replacement heifers—We assumed that two classes of heifers were fed based on annual inventory numbers provided by the U.S. Department of Agriculture, annual cattle inventory report (USDA/NASS, 2009-11). NASS reported the number of heifers for beef cow replacement and the number expected to calve. Younger heifers not expected to calve equaled the difference between these two inventory numbers. Heifers expected to calve were assumed to be 900 pound animals with an average daily dry matter intake of 19.0 pounds during the last 4 months prior to calving (Ensminger et al., 1990).

Heifers not expected to calve were assumed to be 650 pound animals with an average daily dry matter intake of 15.6 pounds (Ensminger et al., 1990). We assumed that both types of heifers were fed at a 20-percent DDGS inclusion rate (Klopfenstein et al., 2008; U.S. Grain Council, 2007). The daily potential amount of DDGS fed to heifers = a weighted average of 3.89 pounds ((19.0 pounds daily dry matter intake x 20 percent inclusion rate x percent of all beef cow replacement heifers expected to calf (3,563/5,835 = 60.7 percent as of January 1, 2007)) + (15.6 pounds daily dry matter intake x 20-percent inclusion rate x percent of all beef cow replacement heifers not expected to calf (2,272/5,835 = 39.3 percent as of January 1, 2007))) x 110 percent to convert to an as fed basis. These animals were assumed to be fed DDGS for 120 days.

Cattle on feed—We assumed that 950 pound steers consumed 22.1 pounds of dry matter per day with a 35-percent DDGS inclusion rate (Klopfenstien et al., 2008; Erickson et al., 2007). The daily amount of DDGS fed to cattle on feed = 8.51 pounds (22.1pounds of daily dry matter intake x 35-percent DDGS inclusion rate) x 110 percent to convert to an as fed basis. The 35-percent inclusion rate represents either wet or dry feeding of distillers' grains with solubles. We assumed that these animals were fed 365 days of the year. The number of cattle fed came from USDA, NASS, 2009-11.

Other cattle—This category included other heifers, steers, and bulls weighing more than 500 pounds not in a feedlot and calves weighing less than 500 pounds. For the heavier animals, we assumed that a 650 pound animal consumed 15.6 pounds of daily dry matter (National Research Council, 2000), and for the calves, we assumed that a 350 pound animal consumed 10 pounds of daily dry matter (Ensminger et al., 1990). The daily potential amount of DDGS fed to other cattle = a weighted average of 2.80 pounds ((15.6 pounds daily dry matter intake x 20-percent inclusion rate x percent of heifers, steers, and bulls over 500 pounds not in a feedlot (48.9 percent as of January 1, 2007)) + (10.0 pounds daily dry matter intake x 20-percent inclusion rate x percent of calves under 500 pounds and not in a feedlot (51.1 percent as of January 1, 2007))) x 110 percent to convert to an as fed basis. We assumed that these animals for both categories were fed 120 days of the year.

Dairy Cattle

Cows—We used a 52.0 pound mid-lactation daily dry matter intake times a 20-percent DDGS inclusion rate to arrive at the daily amount fed (Hutjens,

2008; Schingoethe, 2008). We computed a weighted daily average amount of DDGS fed to account for lactating and dry cows. For example, we assumed that the cows lactated for 305 days and were dry for 60 days per year. We assumed that a 1,500 pound dry cow was fed an average daily dry matter of 27 pounds (average during the dry period) (Hutjens, 2008) with a 10-percent DDGS inclusion rate (Schingoethe, 2008). Daily potential amount of DDGS fed to dairy cows = a weighted average of 10.05 pounds ((52.0 pounds daily dry matter intake x 20-percent inclusion rate x 305/365 days fed per year) + (27.0 pounds average daily dry matter intake x 10-percent inclusion rate x 60/365 days fed per year)) x 110 percent to covert to an as fed basis. The inclusion rates represent either WDGS or DDGS.

Milk cow replacement heifers—We made assumptions on two classes of heifers based on the annual inventory numbers provided by USDA's annual cattle inventory report (USDA/NASS, 2009-11). These findings included the number of heifers for milk cow replacement and the number expected to calve. Younger heifers not expected to calve equaled the difference between these two inventory numbers. Heifers expected to calve were assumed to be 880 pound animals with a daily dry matter intake of 19.4 pounds (Hutjens, 2008). Heifers not expected to calve were assumed to be 660 pound animals with a 15.6-daily dry matter intake. The daily potential amount of DDGS fed to heifers = a weighted average of 3.98 pounds ((19.4 pounds daily dry matter intake x 20-percent inclusion rate x percent of heifers expected to calve as a percent of all milk cow replacement heifers (2,831/4,325 = 65.8 percent as of January 1, 2007) x 110 percent to convert to an as fed basis.) + (15.6 pounds daily dry matter intake x 20-percent inclusion rate x percent of heifers not expected to calve as a percent of all milk cow replacement heifers(1,494/4,325 = 34.2 percent as of January 1, 2007) x 110 percent to convert to an as fed basis)). We assumed these animals were fed DDGS for 120 days per year.

Swine

To calculate the potential amount of DDGS fed to swine in a given crop year, we determined the average quarterly number of breeding swine per crop year and the average quarterly number of market hogs per weight group (USDA/ NASS, *Quarterly Hogs and Pigs* report, various years and quarters). We determined the inventory numbers for each category and assumed that they were fed for 365 days of the year.

Breeding swine—We assumed that each hog in the breeding swine category had two production cycles per year. So we assumed an average weight of 358.3 pounds for a bred gilt, sow, and adult boar with a daily feed intake of 4.2 pounds per day and assumed that they were fed this diet for about 4 months per production cycle, or a total of 8 months per year (Ensminger et al., 1990). We also assumed an average 363.8 pound lactating gilt and sow consumed 11.7 pounds of feed per day for 2 months per production cycle, or 4 months per year (Ensminger et al., 1990). We used a 20-percent DDGS inclusion rate in our calculations. The daily quantity of DDGS fed to breeding swine = a weighted average of 1.34 pounds (4.2 pounds of feed per day x 20-percent DDGS inclusion x 8/12 months) + (11.7 pounds of feed per day x 20-percent DDGS inclusion rate per day x 4/12 months).

Market swine—We calculated the number of market hogs fed based on the average quarterly inventory of market hogs by weight category. For example, we considered the following weight groups:

1) Under 50 pounds market hogs were assumed to have been fed a ration of 2 pounds per day with a 10- percent DDGS inclusion rate.
2) The 50-119 pound market hogs were assumed to have been fed a ration of 4.2 pounds per day with a 20-percent DDGS inclusion rate.
3) The 120-179 pound market hogs were assumed to have been fed a ration of 6.9 pounds per day with a 20-percent DDGS inclusion rate.
4) The 180 pounds and over market hogs were assumed to have been fed a ration of 6.9 pounds per day with a DDGS inclusion rate of 10 percent (Ensminger et al., 1990).

The daily quantity of DDGS fed to market hogs during 2006/07 = a weighted average of 0.68 pounds per day (2 pounds of feed per day x 20-percent DDGS inclusion rate x portion of all market hogs in the under 50 pounds category (20,823/56,688 = 37 percent) + (4.2 pounds of feed fed per day x 20-percent DDGS inclusion rate x portion of all market hogs in the 50-119 pound category (13,958/56,688 = 25 percent) + (6.9 pounds of feed fed per day x 20-percent DDGS inclusion rate x portion of all market hogs in the 120-179 pound category (11,606/56,688 = 20 percent) + (6.9 pounds of feed fed per day x 10-percent DDGS inclusion rate x portion of all market hogs in the 180 pounds and over category (10,301/56,688 = 18 percent).

Poultry

To compute the potential amount of DDGS that poultry could be fed during the corn crop year, we calculated the amount of DDGS fed per day times the days fed per year times the number of birds fed. For the number of birds fed during the crop year, we used an average beginning-of-the-month inventory number for layers and pullets and the number of birds slaughtered for broilers and turkeys (see tables 3 through 7).

Layers—We assumed that layers were fed for 10 months of the year for egg production and 2 months of the year were allocated to molting. During the molting period, they receive nothing but water for 2 weeks, and for the remaining 6 weeks, they receive a maintenance ration (Ensminger et al., 1990). We assumed a DDGS inclusion rate of 12 percent. We assumed that a 60 week old layer consumed 1.65 pounds of feed per week while producing eggs and a maintenance diet of 1.01 pounds per week during molting (Ensminger et al., 1990). The daily amount of DDGS fed to layers = a weighted average of 0.026 pounds ((0.2357 pounds of feed consumption per day x 12-percent DDGS inclusion x 10/12 months) + (0 pounds of feed consumption per day for 2 weeks) + (0.1442 pounds of feed consumption per day x 12-percent DDGS inclusion x 1.5/12 months)).

Pullets—We assumed that pullets were fed a 12-percent DDGS inclusion rate from 2 to 25 weeks, receiving an average of 0.1442 pounds per day for 365 days (Ensminger et al., 1990). The daily amount of DDGS fed to pullets = 0.017 pounds (0.1442 pounds of feed per day x 12-percent DDGS inclusion rate).

Broilers—We assumed that broilers were fed for 8 weeks, but for the first week DDGS were not included in the ration. The grow-finish ration lasted 7 weeks, for which we assumed a 10- percent DDGS inclusion rate. We averaged the cumulative amount fed to male and female broilers for the 7 weeks and divided by 49 days for an average daily amount fed times the DDGS inclusion rate. The daily amount of DDGS fed to broilers = a weighted average of 0.022 pounds ((((11.49 total pounds for male broilers weeks 2-7) + (9.75 total pounds for female broilers weeks 2-7) / 2) / 49days) x 10- percent DDGS inclusion rate in daily feed intake) (Ensminger et al., 1990).

Turkeys—We assumed that turkeys were fed for 6 months (male) and 5 months (female). DDGS were not fed to turkeys for the first 3 weeks. Starting week 4 and each week thereafter, DDGS were included in the ration at a 10-percent inclusion rate. We assumed that during this period male turkeys consumed 114.09 pounds of feed and females consumed 54.3 pounds of feed.

We summed and divided these numbers by 2 to get an average and then divided by 147 days (the number of days fed DDGS). The daily amount of DDGS fed to turkeys = a weighted average of 0.057 pounds ((((114.09 total pounds of feed for males turkeys weeks 4-24 + 54.3 total pounds of feed for female turkeys weeks 4-20) / 2) / 147 days) x 10-percent of DDGS inclusion in daily feed intake) (Ensminger et al., 1990).

APPENDIX B: EFFECTS OF ALTERNATIVE MARKET SHARE ASSUMPTIONS UPON DDGS SUBSTITU TION FOR CORN AND SOYBEAN MEAL, BY TYPE OF LIVESTOCK/POULTRY

Appendix Table 1. Total U.S. DDGS substitution for corn and soybean meal, by livestock/poultry type, crop year 2006/07

	Type of livestock/poultry				
	Beef cattle	Dairy cattle	Swine	Poultry	Total
Estimated:					
Market share (percent)	42.0	43.3	10.4	4.3	100.0
Consumption (mmt)	5.3	5.4	1.3	0.5	12.5
Substitution rate for...[1]					
1 pound of DDGS substitutes for how many pounds of corn:			*Pounds*		
Alternative #1	1.00	0.45	0.89	0.51	0.73
Alternative #2	1.20	0.73	0.70	0.61	0.92
1 pound of DDGS substitutes for how many pounds of soybean meal:					
Alternative #1	0.00	0.55	0.10	0.50	0.27
Alternative #2	0.00	0.63	0.30	0.44	0.32
12.5 mmt of DDGS substitutes for how many mmt of corn:			*mmt*		
Alternative #1	5.3	2.4	1.2	0.3	9.1
Alternative #2	6.3	4.0	0.9	0.3	11.5
12.5 mmt of DDGS substitutes for how many mmt of soybean meal:					
Alternative #1	0.0	3.0	0.1	0.3	3.4
Alternative #2	0.0	3.4	0.4	0.2	4.0
mmt=Million metric tons.					

[1]Converting from pounds to metric tons the substitution rate becomes—Alternative #1: Substitution rate = 1 metric ton of DDGS substitutes for 1 metric ton of feed consisting of 0.73 metric ton of corn and 0.27 metric ton soybean meal. Alternative #2: Substitution rate = 1 metric ton DDGS substitutes for 1.24 metric ton of feed consisting of 0.92 metric ton of corn and 0.32 metric ton of soybean meal.

Source: Renewable Fuels Association market share estimates and tables 10, 11, and 14 from this report.

Appendix Table 2. Total U.S. DDGS substitution for corn and soybean meal, by livestock/poultry type, crop year 2007/08

	Type of livestock/poultry				
	Beef cattle	Dairy cattle	Swine	Poultry	Total
Estimated:					
Market share (percent)	39.3	42.0	13.0	5.7	100.0
Consumption (mmt)	6.9	7.4	2.3	1.0	17.6
Substitution rate for..[1]					
1 pound of DDGS substitutes for how many pounds of corn:			*Pounds*		
Alternative #1	1.00	0.45	0.89	0.51	0.73
Alternative #2	1.20	0.73	0.70	0.61	0.90
1 pound of DDGS substitutes for how many pounds of soybean meal:					
Alternative #1	0.00	0.55	0.10	0.50	0.27
Alternative #2	0.00	0.63	0.30	0.44	0.33
17.6 mmt of DDGS substitutes for how many mmt of corn:			*mmt*		
Alternative #1	6.9	3.3	2.0	0.5	12.8
Alternative #2	8.3	5.4	1.6	0.6	15.9
17.6 mmt of DDGS substitutes for how many mmt of soybean meal:					
Alternative #1	0.0	4.1	0.2	0.5	4.8
Alternative #2	0.0	4.7	0.7	0.4	5.8
mmt=Million metric tons.					

[1]Converting from pounds to metric tons the substitution rate becomes—Alternative #1: Substitution rate = 1 metric ton of DDGS substitutes for 1 metric ton of feed consisting of 0.73 metric ton of corn and 0.27 metric of soybean meal. Alternative #2: Substitution rate = 1 metric of DDGS substitutes for 1.23 metric ton of feed consisting of 0.90 metric ton of corn and 0.33 metric ton of soybean meal.

Source: Renewable Fuels Association market share estimates and tables 10, 11, and 14 from this report.

Appendix Table 3. Total U.S. DDGS substitution for corn and soybean meal, by livestock/poultry type, crop year 2008/09

	Type of livestock/poultry				
	Beef cattle	Dairy cattle	Swine	Poultry	Total
Estimated: Market share (percent) Consumption (mmt)	38.3 8.3	40.3 8.8	14.7 3.2	6.7 1.5	100.0 21.8
Substitution rate for..[1]					
1 pound of DDGS substitutes for how many pounds of corn:			*Poun ds*		
Alternative #1	1.00	0.45	0.89	0.51	0.73
Alternative #2	1.20	0.73	0.70	0.61	0.89
1 pound of DDGS substitutes for how many pounds of soybean meal:					
Alternative #1	0.00	0.55	0.10	0.50	0.27
Alternative #2	0.00	0.63	0.30	0.44	0.33
21.5 mmt of DDGS substitutes for how many mmt of corn:			*mmt*		
Alternative #1	8.3	4.0	2.9	0.8	15.9
Alternative #2	10.0	6.4	2.3	0.9	19.6
21.5 mmt of DDGS substitutes for how many mmt of soybean meal:					
Alternative #1	0.0	4.8	0.3	0.7	5.9
Alternative #2	0.0	5.5	1.0	0.7	7.1
mmt=Million metric tons.					

[1]Converting from pounds to metric tons the substitution rate becomes—Alternative #1: Substitution rate = 1 metric ton of DDGS substitutes for 1 metric ton of feed consisting of 0.73 metric ton of corn and 0.27 metric of soybean meal. Alternative #2: Substitution rate = 1 metric of DDGS substitutes for 1.22 metric ton of feed consisting of 0.89 metric ton of corn and 0.33 metric ton of soybean meal.

Source: Renewable Fuels Association market share estimates and tables 10, 11, and 14 from this report.

Appendix Table 4. Total U.S. DDGS substitution for corn and soybean meal, by livestock/poultry type, crop year 2009/10

	Type of livestock/poultry				
	Beef cattle	Dairy cattle	Swine	Poultry	Total
Estimated:					
Market share (percent)	40.4	39.4	11.8	8.4	100.0
Consumption (mmt)	10.3	10.0	3.0	2.1	25.5
Substitution rate for..[1]					
1 pound of DDGS substitutes for how many pounds of corn:			*Pounds*		
Alternative #1	1.00	0.45	0.89	0.51	0.73
Alternative #2	1.20	0.73	0.70	0.61	0.90
1 pound of DDGS substitutes for how many pounds of soybean meal:					
Alternative #1	0.00	0.55	0.10	0.50	0.27
Alternative #2	0.00	0.63	0.30	0.44	0.32
25.5 mmt of DDGS substitutes for how many mmt of corn:			*mmt*		
Alternative #1	10.3	4.5	2.7	1.1	18.6
Alternative #2	12.3	7.3	2.1	1.3	23.1
25.5 mmt of DDGS substitutes for how many mmt of soybean meal:					
Alternative #1	0.0	5.5	0.3	1.1	6.9
Alternative #2	0.0	6.3	0.9	0.9	8.2
mmt=Million metric tons.					

[1]Converting from pounds to metric tons the substitution rate becomes—Alternative #1: Substitution rate = 1 metric ton of DDGS substitutes for 1.00 metric ton of feed consisting of 0.73 metric ton of corn and 0.27 metric ton of soybean meal. Alternative #2: Substitution rate = 1 metric ton of DDGS substitutes for 1.22 metric ton of feed consisting of 0.90 metric ton of corn and 0.32 metric ton of soybean meal.

Source: Renewable Fuels Association market share estimates and tables 10, 11, and 14 from this report.

Appendix Table 5. Total U.S. DDGS substitution for corn and soybean meal, by livestock/poultry type, crop year 2010/11

	Type of livestock/poultry				
	Beef cattle	Dairy cattle	Swine	Poultry	Total
Estimated: Market share (percent) Consumption (mmt)	41.4 12.0	39.4 11.5	10.1 2.9	9.1 2.6	100.0 29.1
Substitution rate for..[1]					
1 pound of DDGS substitutes for how many pounds of corn:			*Pounds*		
Alternative #1	1.00	0.45	0.89	0.51	0.73
Alternative #2	1.20	0.73	0.70	0.61	0.91
1 pound of DDGS substitutes for how many pounds of soybean meal:					
Alternative #1	0.00	0.55	0.10	0.50	0.27
Alternative #2	0.00	0.63	0.30	0.44	0.32
29.1mmt of DDGS substitutes for how many mmt of corn:			*mmt*		
Alternative #1	12.0	5.2	2.6	1.4	21.2
Alternative #2	14.4	8.4	2.1	1.6	26.5
29.1 mmt of DDGS substitutes for how many mmt of soybean meal:					
Alternative #1	0.0	6.3	0.3	1.3	7.9
Alternative #2	0.0	7.2	0.9	1.2	9.3
mmt=Million metric tons.					

[1]Converting from pounds to metric tons the substitution rate becomes—Alternative #1: Substitution rate = 1 metric ton of DDGS substitutes for 1.00 metric ton of feed consisting of 0.73 metric ton of corn and 0.27 metric ton of soybean meal. Alternative #2: Substitution rate = 1 metric ton of DDGS substitutes for 1.22 metric ton of feed consisting of 0.91 metric ton of corn and 0.32 metric ton of soybean meal.

Source: Renewable Fuels Association market share estimates and tables 10, 11, and 14 from this report.

Appendix Table 6. Total U.S. DDGS substitution for corn and soybean meal, by livestock/poultry type, crop year 2006/07

	Type of livestock/poultry				
	Beef cattle	Dairy cattle	Swine	Poultry	Total
Estimated:					
Market share (percent)	45.2	42.8	6.0	6.0	100.0
Consumption (mmt)	5.6	5.3	0.8	0.8	12.5
Substitution rate for..[1]					
1 pound of DDGS substitutes for how many pounds of corn:			*Pounds*		
Alternative #1	1.00	0.45	0.89	0.51	0.73
Alternative #2	1.20	0.73	0.70	0.61	0.93
1 pound of DDGS substitutes for how many pounds of soybean meal:					
Alternative #1	0.00	0.55	0.10	0.50	0.27
Alternative #2	0.00	0.63	0.30	0.44	0.31
12.5 mmt of DDGS substitutes for how many mmt of corn:			*mmt*		
Alternative #1	5.6	2.4	0.7	0.4	9.0
Alternative #2	6.7	3.9	0.5	0.5	11.6
12.5 mmt of DDGS substitutes for how many mmt of soybean meal:					
Alternative #1	0.0	2.9	0.1	0.4	3.4
Alternative #2	0.0	3.3	0.2	0.3	3.9
mmt=Million metric tons.					

[1]Converting from pounds to metric tons the substitution rate becomes—Alternative #1: Substitution rate = 1 metric ton of DDGS substitutes for 1 metric ton of feed consisting of 0.73 metric ton of corn and 0.27 metric ton of soybean meal. Alternative #2: Substitution rate = 1 metric ton of DDGS substitutes for 1.24 metric ton of feed consisting of 0.93 metric ton of corn and 0.31 metric ton of soybean meal.

Source: Wisner (2011) market share estimates and tables 10, 11, and 14 from this report.

Appendix Table 7. Total U.S. DDGS substitution for corn and soybean meal, by livestock/poultry type, crop year 2007/08

	Type of livestock/poultry				
	Beef cattle	Dairy cattle	Swine	Poultry	Total
Estimated: Market share (percent) Consumption (mmt)	47.6 8.4	40.4 7.0	6.0 1.1	6.0 1.1	100.0 17.6
Substitution rate for..[1]					
1 pound of DDGS substitutes for how many pounds of corn:			*Pounds*		
Alternative #1	1.00	0.45	0.89	0.51	0.74
Alternative #2	1.20	0.73	0.70	0.61	0.94
1 pound of DDGS substitutes for how many pounds of soybean meal:					
Alternative #1	0.00	0.55	0.10	0.50	0.26
Alternative #2	0.00	0.63	0.30	0.44	0.30
17.6 mmt of DDGS substitutes for how many mmt of corn:			*mmt*		
Alternative #1	8.4	3.2	0.9	1.3	13.1
Alternative #2	10.0	5.1	0.7	1.6	16.6
17.6 mmt of DDGS substitutes for how many mmt of soybean meal:					
Alternative #1	0.0	3.9	0.1	1.3	4.5
Alternative #2	0.0	4.4	0.3	1.1	5.2
mmt=Million metric tons.					

[1]Converting from pounds to metric tons the substitution rate becomes—Alternative #1: Substitution rate = 1 metric ton of DDGS substitutes for 1 metric ton of feed consisting of 0.74 metric ton of corn and 0.26 metric ton of soybean meal. Alternative #2: Substitution rate = 1 metric ton of DDGS substitutes for 1.24 metric ton of feed consisting of 0.94 metric ton of corn and 0.30 metric ton of soybean meal.

Source: Wisner (2011) market share estimates and tables 10, 11, and 14 from this report.

Appendix Table 8. Total U.S. DDGS substitution for corn and soybean meal, by livestock/poultry type, crop year 2008/09

	Type of livestock/poultry				
	Beef cattle	Dairy cattle	Swine	Poultry	Total
Estimated:					
Market share (percent)	50.0	38.0	6.0	6.0	100.0
Consumption (mmt)	10.9	8.3	1.3	1.3	21.8
Substitution rate for..[1]					
1 pound of DDGS substitutes for how many pounds of corn:			*Pounds*		
Alternative #1	1.00	0.45	0.89	0.51	0.76
Alternative #2	1.20	0.73	0.70	0.61	0.95
1 pound of DDGS substitutes for how many pounds of soybean meal:					
Alternative #1	0.00	0.55	0.10	0.50	0.24
Alternative #2	0.00	0.63	0.30	0.44	0.28
21.8 mmt of DDGS substitutes for how many mmt of corn:			*mmt*		
Alternative #1	10.9	3.7	1.2	0.7	16.5
Alternative #2	13.0	6.1	0.9	0.8	20.8
21.8 mmt of DDGS substitutes for how many mmt of soybean meal:					
Alternative #1	0.0	4.6	0.1	0.7	5.3
Alternative #2	0.0	5.2	0.4	0.6	6.2
mmt=Million metric tons.					

[1]Converting from pounds to metric tons the substitution rate becomes—Alternative #1: Substitution rate = 1 metric ton of DDGS substitutes for 1 metric ton of feed consisting of 0.76 metric ton of corn and 0.24 metric ton of soybean meal. Alternative #2: Substitution rate = 1 metric ton of DDGS substitutes for 1.23 metric ton of feed consisting of 0.95 metric ton of corn and 0.28 metric ton of soybean meal.

Source: Wisner (2011) market share estimates and tables 10, 11, and 14 from this report.

Appendix Table 9. Total U.S. DDGS substitution for corn and soybean meal, by livestock/poultry type, crop year 2009/10

	Type of livestock/poultry				
	Beef cattle	Dairy cattle	Swine	Poultry	Total
Estimated: Market share (percent) Consumption (mmt)	52.4 13.4	35.6 9.1	6.0 1.5	6.0 1.5	100.0 25.5
Substitution rate for..[1]					
1 pound of DDGS substitutes for how many pounds of corn:			*Pounds*		
Alternative #1	1.00	0.45	0.89	0.51	0.77
Alternative #2	1.20	0.73	0.70	0.61	0.96
1 pound of DDGS substitutes for how many pounds of soybean meal:					
Alternative #1	0.00	0.55	0.10	0.50	0.23
Alternative #2	0.00	0.63	0.30	0.44	0.27
25.5 mmt of DDGS substitutes for how many mmt of corn:			*mmt*		
Alternative #1	13.4	4.1	1.4	0.8	19.6
Alternative #2	16.0	6.6	1.1	0.9	24.6
25.5 mmt of DDGS substitutes for how many mmt of soybean meal:					
Alternative #1	0.0	5.0	0.2	0.8	5.9
Alternative #2	0.0	5.7	0.5	0.7	6.9
mmt=Million metric tons.					

[1]Converting from pounds to metric tons the substitution rate becomes—Alternative #1: Substitution rate = 1 metric ton of DDGS substitutes for 1 metric ton of feed consisting of 0.77 metric ton of corn and 0.23 metric ton of soybean meal. Alternative #2: Substitution rate = 1 metric ton of DDGS substitutes for 1.23 metric ton of feed consisting of 0.96 metric ton of corn and 0.27 metric ton of soybean meal.

Source: Wisner (2011) market share estimates and tables 10, 11, and 14 from this report.

Appendix Table 10. Total U.S. DDGS substitution for corn and soybean meal, by livestock/poultry type, crop year 2010/11

	Type of livestock/poultry				
	Beef cattle	Dairy cattle	Swine	Poultry	Total
Estimated:					
Market share (percent)	53.4	34.1	6.9	5.6	100.0
Consumption (mmt)	15.5	9.9	2.0	1.6	29.1
Substitution rate for...[1]					
1 pound of DDGS substitutes for how many pounds of corn:			*Pounds*		
Alternative #1	1.00	0.45	0.89	0.51	0.78
Alternative #2	1.20	0.73	0.70	0.61	0.97
1 pound of DDGS substitutes for how many pounds of soybean meal:					
Alternative #1	0.00	0.55	0.10	0.50	0.22
Alternative #2	0.00	0.63	0.30	0.44	0.26
29.1 mmt of DDGS substitutes for how many mmt of corn:			*mmt*		
Alternative #1	15.5	4.5	1.8	0.8	22.6
Alternative #2	18.6	7.3	1.4	1.0	28.2
29.1 mmt of DDGS substitutes for how many mmt of soybean meal:					
Alternative #1	0.0	5.5	0.2	0.8	6.5
Alternative #2	0.0	6.3	0.6	0.7	7.6
mmt=Million metric tons.					

[1]Converting from pounds to metric tons the substitution rate becomes—Alternative #1: Substitution rate = 1 metric ton of DDGS substitutes for 1 metric ton of feed consisting of 0.78 metric ton of corn and 0.22 metric ton of soybean meal. Alternative #2: Substitution rate = 1 metric ton of DDGS substitutes for 1.23 metric ton of feed consisting of .97 metric ton of corn and 0.26 metric ton of soybean meal.

Source: Wisner (2011) market share estimates and tables 10, 11, and 14 from this report.

End Notes

[1] In addition to Government policies and higher energy prices in the long run, ethanol production growth depends partly on feedstock costs and changes in technological advancements.

[2] Ethanol production numbers are expressed in terms of the corn crop year (September-August).

[3] Monthly ethanol production data were converted to corn crop years from the usual ethanol calendar year (U.S. Energy Information Administration, 2010).

[4] 1 metric ton = 2,204.622 pounds.

[5] Distillers' grains refer to coproducts generated by dry-mill ethanol plants, including distillers' wet grains (DWG), distillers' dried grains (DDG), distillers' wet grains with solubles (DWGS), distillers' dried grains with solubles (DDGS), and condensed distillers' solubles (CDS). Unless otherwise specified, distillers' grains will refer to distillers' dried grains with solubles (DDGS). See Hoffman and Baker (2010) for additional details about the market for distillers' grains. The Renewable Fuels Association (2011) estimates that 61 percent of distillers' grains products were sold in dry form and 39 percent in wet form for calendar year 2010. The USDA report *Ethanol Co-Products Used in Livestock Feed*, summarizes the average annual quantities of these different coproducts fed by selected livestock producers (USDA/NASS, 2007). Since these statistics could not be used to compute the total amount of these feed products consumed by type of livestock/poultry and since there are no reliable statistics on the production of different ethanol coproducts, we had to simplify the estimation process for each coproduct's substitution potential for corn and soybean meal. We converted all coproducts into what we refer to throughout this report as DDGS. Dry matter basis is used to estimate the substitution of DDGS for corn and soybean meal.

[6] New technology will affect the composition of DDGS, and may change the substitution rates for corn and protein meal.

[7] A ruminant is an animal with a rumen compartment in its stomach (e.g., cattle, sheep, goats, deer, elk, and bison). Unlike animals with a single compartment stomach (monogastric), ruminants are able to convert cellulose in its various forms into starches and sugars, which can then be metabolized.

[8] The precise color preferred by some animal feeders may also depend on the livestock/poultry being fed. Also, color may be a function of type of feedstock used, not always an indicator of protein damage.

[9] Dry matter basis means an adjustment is made for the moisture content, compared with an "as fed" or "as is" basis, where no adjustment is made for moisture content.

[10] Polioencephalomalacia (PEM) is a neurologic disease of ruminants. The term PEM denotes a lesion with certain gross and microscopic features. Historically, PEM has been associated with a thiamine (Vitamin B1) deficiency but more recently has been associated with high sulfur intake.

[11] Recent feeding studies indicate that this limit in diets is 0.53 percent (Erickson et al., 2005).

[12] Excessive phosphorus released into the environment (e.g., through animal waste) may contribute to eutrophication in the Nation's lakes and streams (Rausch and Belyea, 2005). Regulations for disposal of animal wastes are becoming increasingly stringent depending on phosphorus content. Since phosphorus content from DDGS is high relative to corn and to requirements of most ruminants, high phosphorus levels in diets can increase phosphorus in animal wastes. Disposal of these wastes must comply with established regulations.

[13] Diet formulation with specific nutrient requirements by livestock/ poultry species along with a nutrient value for selected feed ingredients can be obtained from the National Research Council (1994, 1998, 2000, and 2001) for poultry, swine, beef, and dairy, respectively.

[14] This brief review is intended to provide a basic understanding of feeding DDGS to different types of livestock/poultry and does not cover all possible references and information about this feeding. For more information about the physiology of each type of livestock/poultry or why they might consume different amounts of distillers' spent grains, see Tisch (2006).

[15] Acidosis can be an acute or subacute condition in rumen livestock. Acute acidosis occurs with rapid grain overload and may result in liver abscesses, severe illness, and even death. If these problems are prolonged, the low rumenal pH may damage the rumenal wall and reduce absorption capacity. Thus, even animals that survive may develop chronic conditions. A major symptom of subacute acidosis is reduced feed intake, making it difficult to diagnose and separate from other problems or events that reduce feed consumption, such as digestibility of the grain. Other symptoms include lethargy, diarrhea, panting, excessive salivation, kicking at the belly, and general signs of discomfort and stress.

[16] This study was a joint effort between USDA's National Agricultural Statistics Service and the Nebraska Corn Development, Utilization & Marketing Board (an agency of the State of Nebraska). Approximately 9,400 livestock operations in Illinois, Indiana, Iowa, Kansas, Michigan, Minnesota, Missouri, Nebraska, North Dakota, Ohio, South Dakota, and Wisconsin were contacted by mail in February 2007 with a second mailing 2 weeks later and telephone follow-up in March. Caution should be used while interpreting the results of the survey due to its regional nature.

[17] For more information about feeding DDGS to poultry, see Bregendahl (2008).

[18] See Appendix A for details on estimating daily potential consumption of DDGS by livestock/poultry.

[19] A least-cost feed ration model (Dahlke and Lawrence, 2008) is another method to compute potential DDGS consumption. However, this option and its massive data requirements for national estimates are beyond the scope of this study.

[20] We assume a potential inclusion rate that is recommended in the literature. This rate may be the low or mid-point of any suggested maximum potential inclusion rate. In doing so, we realize that actual inclusion in any given diet may be lower or higher than our assumed amounts due to local feed availability or relative prices of feed ingredients.

[21] These estimates assume a composition of DDGS from dry-mill corn ethanol plants without new technology applied to the production process. Co-products from new technology dry-mill plants are discussed by Hoffman and Baker (2010).

[22] Dooley (2008) computed an adoption rate based on the livestock/poultry herd/flock size and its ability to consume a truckload of DDGS. While this is a plausible approach, it raises questions about smaller dairy or beef cow farms/ranches that consume DDGS in less than a truckload lot. For example, many smaller dairy farms can consume DDGS by having less than truckload quantities mixed and delivered with their typical concentrates. Fox (2008) used adoption rates derived from the 2007 USDA, NASS survey and applied these somewhat dated adoption rates to estimates of potential DDGS consumption for calendar year 2008.

[23] If we assume 15 billion gallons of corn-based ethanol are produced, corn use for dry-mill fuel ethanol would total 5.17 billion bushels (2.7 gallons ethanol per bushel of corn and 93 percent of ethanol produced from dry-mill plants), or 41.0 mmt, 1 mmt from beverage distilleries, and .5 mmt from imports equal a total supply of about 42.5 mmt.

[24] Feeding studies at South Dakota State University have also shown that DDGS and soy hulls can replace all soybean meal (Staff, 2011).

[25] One survey provided DDGS consumption estimates for beef cattle, fed cattle, dairy cattle, and swine for selected States in calendar year 2006 (USDA/NASS, 2007). However, estimates

for poultry were not made. This survey focused on a selected number of Midwestern States, but lacks a national scope.

[26] The greatest potential for increased DDGS feeding remains with beef cattle. For example, based on an assumed production level of 15 billion gallons of ethanol and estimated domestic consumption levels of 31.5 mmt, the market share by type of livestock/poultry is estimated to be 55.7 percent for beef cattle, 27.2 percent for dairy cattle, 10.5 percent for swine, and 6.6 percent for poultry (table 13).

[27] Beef and dairy cattle received their full remaining potential for each crop year, but swine was reduced by 25 percent and poultry by 50 percent for each crop year because of slower adoption as reported by the industry (see earlier discussion on DDGS inclusion in swine and poultry diets).

[28] For example, Shurson (March 2009) reported that some hogs were fed at a DDGS inclusion rate of over 30 percent during 2008 and 2009, when corn and soybean meal prices were high and prices of DDGS were relatively low. Furthermore, some turkeys were fed at a DDGS inclusion rate of 20-30 percent between 2008 and 2009. In both examples, the inclusion rates are higher than those used in our analysis.

[29] CHS Inc. (http://www.chsinc. com/) is a Fortune 500 company that is diversified in energy, grains, and food. It is owned by farmers, ranchers, cooperatives, and preferred stockholders covering a geographic area from the Great Lakes to the Pacific Northwest and from the Canadian border to Texas. CHS provides products and services ranging from grain marketing to food processing, operates petroleum refineries/pipelines and, through a broad range of working partnerships, markets and distributes Cenex® brand energy products and renewable fuels, along with agronomic inputs and livestock feed. CHS Inc. is a prominent marketer of DDGS in both the domestic and export market.

[30] Note that some commercial feeders are finding ways to use higher levels of DDGS that have not been tested by university feeding studies or peer reviewed publications. For example, there can be a complete replacement of corn with WDGS and a corn stalk or straw mixture for beef cattle.

[31] These finding apply to DDGS fed in the United States. DDGS exports may have the same substitution effect, assuming that all substitution rates and market shares by type of livestock/poultry were similar to the United States. However, this is probably not the case. Thus, substitution rates for each country would have to be analyzed to assess U.S. DDGS imports to substitute for corn and soybean meal in that country.

[32] Projections from the September 12, 2011WASDE report show that 5.02 billion bushels of corn will be used in the production of fuel ethanol for crop year 2010/11. Note that not all corn used to produce fuel ethanol is consumed by corn dry-mill plants. Based on an update of Hoffman and Baker's (2010) estimates, 444.0 million bushels, or 8.8 percent of this total, will be used for ethanol by corn wet-mill plants and 4,576 million bushels, or 91.2 percent of this total, will be used for ethanol by corn dry-mill plants.

[33] The amount of corn used to produce these DDGS in 2006/07 was 40 mmt ((12.5 mmt/0.3125) = 40 mmt)). A 17.5 pound yield of DDGS from a 56 pound bushel of corn used to produce ethanol from a dry-mill corn ethanol plant (17.5 pounds /56 pounds) equaled 31.25 percent of the corn's original weight. DDGS used in the 2006/07 domestic feed market (12.5 mmt (table 11)) substituted for an estimated 15.1 mmt of corn and soybean meal (see table 15), the equivalent of about 37.8 percent (by weight) of the corn used in the associated ethanol production process (15.1 mmt / 40 mmt) = 37.8 percent.

[34] The amount of corn used to produce these DDGS in 2010/11 was estimated to be 93.1 mmt ((29.1 mmt / .3125) = 93.1 mmt). DDGS estimated to be used in the 2010/11 domestic feed

market, (29.1 mmt (table 13)), substituted for an estimated 35.3 mmt of corn and soybean meal or the equivalent of 37.8 percent (weight basis) of the corn used in the associated ethanol production process (35.3 mmt / 93.1 mmt) = 37.8 percent.

[35] Assuming production of corn-based ethanol equals 15 billion gallons, corn use for this dry-mill fuel ethanol production could total 5.17 billion bushels (2.7 gallons of ethanol per bushel of corn with 93 percent of ethanol produced from dry-mill plants) which would produce about 41.0 mmt plus 1 mmt from beverage distilleries and .5 mmt from imports = about 42.5 mmt supply. Assuming a 25-percent export share domestic consumption would equal 31.5 mmt. The amount of corn used to produce these DDGS (assuming 15 billion gallon ethanol production from corn) is 100.8 mmt ((31.5 mmt / .3125) = 100.8 mmt). DDGS estimated to be used in this domestic feed market, 31.5 mmt, substituted for an estimated 38.1 mmt of corn and soybean meal, or the equivalent of 37.8 percent (weight basis) of the corn used in the associated ethanol production process (38.1 mmt / 100.8 mmt) = 37.8 percent.

[36] Changing numbers of livestock/ poultry being fed and the relative feed costs of the individual ingredients are contributing factors to changes in the amounts of feedstuffs being fed. For example, tables 3-7 reflect a general decline in the number of beef and dairy cattle being fed over the past 5 years (2006/07 through 2010/11). Also, the price of DDGS has generally been lower than prices for corn and soybean meal. A more detailed analysis of the factors explaining feed use is not part of this report.

In: Distillers' Dried Grains
Editor: Justin N. Schuber

ISBN 978-1-62417-683-8

Chapter 3

China's Market for Distillers Dried Grains and the Key Influences on Its Longer Run Potential*

Michael Jewison and Fred Gale

Abstract

The expansion of corn-based ethanol production in the United States yields a large volume of residual co-products called distillers dried grain with solubles (DDGS). Most DDGS are utilized in the domestic U. S. market, but Chinese importers seeking raw materials for animal feed have emerged as a significant export market. High feed prices and favorable tax treatment within China stimulated a surge of imports of U. S. DDGS during 2009-11. China's potential demand for U. S. DDGS depends on various factors that include the price of corn, Chinese policy, and the availability and price of other substitute feed ingredients, such as the byproducts of grain processing in China (residual products from alcohol production). Demand is robust, but slower growth in the U. S. supply of DDGS and uncertainties about Chinese policy may constrain growth in exports to China.

* This is an edited, reformatted and augmented version of United States Department of Agriculture, A Report from the Economic Research Service, Publication No. FDS-12g-01, dated August 2012.

INTRODUCTION

The rapid expansion of corn use in U. S. ethanol production is well known. Less well known, but also critical to the economics of the industry, are the supply and demand for the residual co-products from ethanol production, known as distillers dried grains with solubles (DDGS). Robust export demand from China for DDGS can simultaneously raise the value to U. S. ethanol producers and satisfy the Chinese livestock industry's hunger for low-cost feed ingredients.

This report examines the recent trends in U. S. -China trade in DDGS and analyzes factors affecting the competitiveness of U. S. DDGS in the Chinese market. The report discusses the role of U. S. DDGS in meeting the nutritional needs of Chinese livestock at minimum cost, the scarcity of feed materials in China, trade policy, quality issues, and resistance to imports from Chinese suppliers of competing feed ingredients.

BACKGROUND: THE U.S. DDGS INDUSTRY AND MARKET

DDGS are the primary co-products in the production of ethanol using the dry-mill process (see box, "What Are DDGS?"). Approximately 31 percent of each U. S. bushel of corn used in dry-mill ethanol production is left over as DDGS. During the 2010/11 corn marketing year (September-August) U. S. corn dry-mill ethanol producers used 116. 2 million metric tons (mmt) of corn to produce 12. 4 billion gallons of fuel ethanol, resulting in the simultaneous production of 36. 3 mmt of DDGS, which can account for, on average, 20 percent of the typical ethanol plant's total revenue. [1] An additional 0. 9 mmt of DDGS were produced by U. S. beverage distilleries in 2010/11. Approximately 91. 1 percent of corn ethanol is produced by dry mills; the remainder is from wet mills. The co-products from wet mills include corn gluten feed and corn gluten meal, but not DDGS (Hoffman and Baker, 2010).

DDGS have emerged to become a useful feed ingredient and an important source of supplementary income for dry-mill ethanol producers. In the United States, DDGS are used primarily for dairy and beef feeding. All livestock and poultry can use the nutrients from DDGS, but beef and dairy cattle (ruminants) can use them more readily than hogs and poultry (monogastrics) (Hoffman and Baker, 2011). DDGS are cheaper when incorporated wet (distillers wet grains (DWG) contain 65-percent moisture), but higher transport costs and shorter shelf life can be problematic. However, DWG actually improve feed quality in the beef and dairy sector as a result of most rations having low moisture content.

What Are DDGS?

The term "distillers grains" refers to co-products generated by dry-mill ethanol plants, including distillers wet grains (DWG), distillers dried grains (DDG), distillers wet grains with solubles (DWGS), distillers dried grains with solubles (DDGS), and condensed distillers solubles (CDS). Unless otherwise specified for the remainder of this report, the term distillers grains will mean distillers dried grains with solubles (DDGS).

Distillers dried grains with solubles (DDGS) are the primary residual co-product after the starch portion of corn is fermented to produce ethanol. After the corn has been fermented, alcohol is removed by distillation, with the remaining residues then dried. Typically, three types of residual co-products are produced via this process: distillers dried grains, distillers dried solubles, and distillers dried grains with solubles.

Once the fermented corn is distilled, the soluble portion of the residual co-products is condensed by evaporation to make distillers solubles. The solid material remaining from the fermentation residues is distillers grains, and both co-products are dried to produce distillers dried grains (DDG) and distillers dried solubles (DDS), with the former having a higher concentration of nutrients than the latter. In the United States, ethanol plants blend and dry DDS and DDG to get DDGS. The estimate for U. S. production is therefore the total of all three products.

China: An Important Market for U.S. DDGS Exports

Most DDGS produced in the United States are used domestically in feed rations for cattle or other livestock. However, feed manufacturers in other countries have shown interest in U. S. DDGS as an economical feed ingredient. China, in particular, with its rising demand for livestock feed and escalating feed costs, has emerged as a significant market. Strong Chinese imports of U. S. DDGS help feed mills in China meet robust feed demand while simultaneously supporting the value of DDGS for the U. S. ethanol industry. Strong markets for DDGS are important to the continued profitability of ethanol production, which in turn supports the demand for U. S. corn.

Exports have absorbed 19 to 25 percent of the U. S. DDGS supply in recent years. U. S. exports of DDGS grew from less than 2. 0 mmt in 2006/07 to over 8. 0 mmt in 2010/11.[2] China was the leading destination of exports during 2009/10, with sales reaching 2. 2 mmt. Exports to China fell to 1. 6 mmt in 2010/11 due to an increase in U. S. prices and the launch of a Chinese antidumping investigation against U. S. DDGS during 2011. Still, during 2010/11, China was a close second behind Mexico as the top destination for U. S. DDGS exports, accounting for 19 percent of the total (figure 1).

DDGS Potential Role in China's Feed Industry

China's demand for DDGS is driven by growth and structural change in its livestock sector. China already is one of the leading producers of animal protein, with meat output in 2010 totaling 79. 3 mmt, including 50. 7 mmt of pork, 16. 0 mmt of poultry meat, 6. 5 mmt of beef, and 4. 0 mmt of mutton.

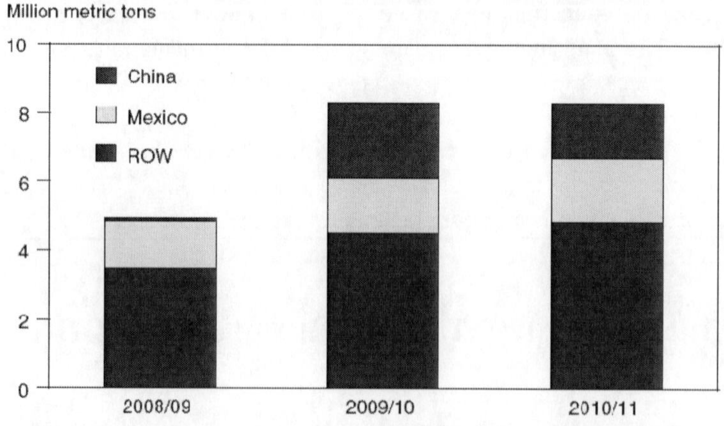

DDGS = Distillers dried grains with solubles.ROW = Rest of world.
Source: USDA, Economic Research Service analysis of U.S. Customs data.

Figure 1. U.S. DDGS exports to China surged in 2009/10.

China also produced 27. 6 mmt of eggs, 37. 5 mmt of milk, and 53. 7 mmt of fish, shellfish, and other aquatic products. All of these activities use commercial feed, and all are expected to grow substantially. The country's 12th

5-year plan for the livestock industry anticipates that livestock production will continue growing to reach 85. 0 mmt of meat, 29. 0 mmt of eggs, and 50. 0 mmt of milk by 2015 (table 1). The 5-year plan for fisheries also forecasts aggressive growth in the production of aquatic products, reaching 60. 0 mmt of production over the same period.[3]

China's imported DDGS are purchased mainly by commercial feed compounders who combine various ingredients to meet the nutritional needs of livestock at minimum cost. Traditionally, Chinese animals and poultry were fed household and processing wastes, crop byproducts, vines, tubers and forages, but the use of feed grains, oilseed meals, and feed additives has increased rapidly since the 1980s. China's commercial feed production rose from under 36. 0 mmt in 1991 to 169. 0 mmt in 2011 (figure 2). China's 5-year plan anticipates that commercial feed production will grow to 200. 0 mmt in 2015. The projected annual increase of 4. 7 percent (7. 6 mmt) is much faster than the projected growth in livestock and poultry output. This rapid growth reflects a shift from traditional feed sources to commercial feed in China's livestock industry.

Table 1. China's 2011-2015 5-year plan targets for livestock and feed

Item	2010 actual	2015 target	Compound annual growth
	Million metric tons		*Percent*
Meat	79.3	85.0	1.4
Eggs	27.6	29.0	1.0
Milk	37.5	50.0	5.9
Aquatic products	53.7	60.0	2.2
Feed	162.0	200.0	4.3

Source: China National Bureau of Statistics and 12th 5-Year Plans for Livestock, Fisheries, and Feed.

As China's demand for feed expands, the country's limited endowment of land and water resources constrains its capacity to expand feed grain production. Supply is also constrained by China's tariff and nontariff measures restricting corn imports. As demand outpaces supply, domestic prices of corn, other grains and oilseed meals are rising (Gale et al., 2012). Feed companies and livestock producers are seeking alternative ingredients that can reduce feed production costs (Wang, 2011).

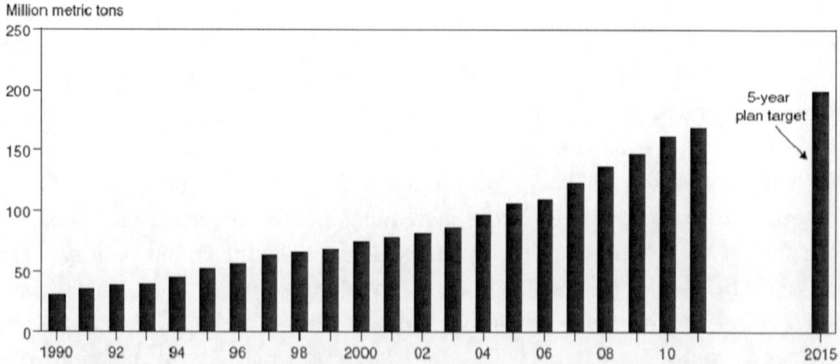

Source: USDA, Economic Research Service analysis of data from China Feed Industry Association.

Figure 2. China manufactured feed output, 1990-2015.

Use of imported DDGS in the Chinese feed industry is still in an early stage. Feed compounders mix a variety of ingredients to produce feed products containing needed energy, protein, amino acids, vitamins, and trace elements at minimum cost. Corn and soybean meal are the predominant ingredients for commercial livestock feed in China, but other major ingredients include wheat and rice bran and hulls; other corn processing byproducts; fish meal; cottonseed and rapeseed meals; meat, blood and bone meals; and meat processing waste. In 2010, the Ministry of Agriculture estimated that 134 mmt of various raw materials were used by the feed industry, with feed use of DDGS estimated at 2. 5 mmt (China Ministry of Agriculture Feed Office, 2011).

The rising domestic price of corn has spurred Chinese feed compounders to seek alternative raw materials like DDGS that can add energy, protein, B vitamins, phosphorus, and trace elements to animal feed formulations at low cost. News reports and industry sources indicate that Chinese purchasers view DDGS primarily as a substitute for corn (China Feed Trade, 2011; China Aquaculture Network, 2010). [4] China has a long tradition of feeding residue from alcohol distilling to swine, but the practice was traditionally confined to farms near distilleries (Coffing et al., 1991). The launch of Chinese grain-based ethanol production in 2004 yielded a domestic DDGS supply used mainly in northern China. Imports of U. S. DDGS increased as Chinese feed prices began rising in recent years, and numerous feed mills in coastal areas began importing DDGS (*Nongcai Baodian*, 2011; China Corn Net, 2011).

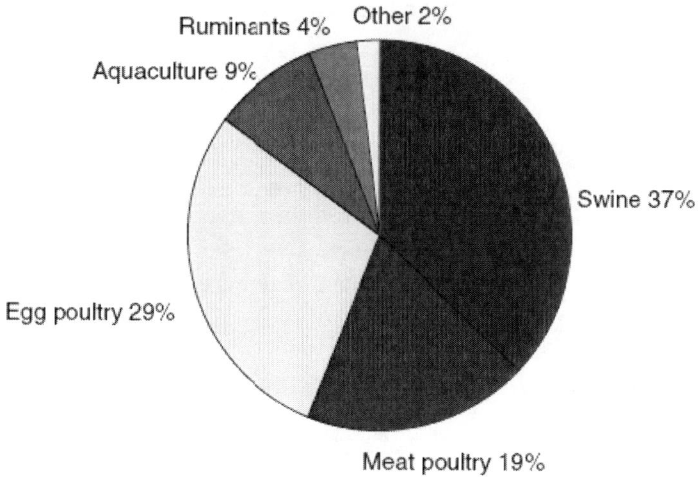

Source: USDA, Economic Research Service analysis of data from China Ministry of Agriculture Feed Industry Office, 2011.

Figure 3. Livestock sector feed output, 2010.

DDGS can be included in the diets of a variety of livestock and poultry species to varying degrees. In the United States, DDGS are used predominantly for feeding cattle. In China, DDGS use is in an experimental stage and is currently in use primarily in feed for poultry and swine (*Nongcai Baodian*, 2011). Chinese feed industry statistics show that nearly half of feed manufactured in 2010 was destined for poultry (mainly chickens and ducks) that produced either eggs or meat, while 37 percent was fed to swine (see figure 3).[5] Aquaculture (fish, shrimp, and other shellfish) accounted for 9 percent of China's manufactured feed. Ruminants (mainly cattle, sheep, and goats) accounted for only 4 percent of manufactured feed output; these animals consume mainly crop residues and forages in China.

The predominance of monogastrics like poultry and hogs in China limits use of DDGS since these species cannot easily digest the large amounts of fiber in DDGS (*Nongcai Baodian*, 2011). In China, DDGS generally is limited to 10-12 percent of hog feed rations and 5-10 percent of poultry rations, but the large number of hogs and poultry in China still represent a large potential market for DDGS. A shift toward large-scale commercial dairy farms following China's 2008 melamine milk-adulteration incident could create more demand for U. S. DDGS for Chinese dairy feed (tables 2a and 2b).

Prices and Other Determinants of Competitiveness

Since cost is a major consideration for Chinese buyers, China's demand for imported DDGS depends largely on the price of DDGS and the price of substitutes. DDGS are a relatively low-cost feed ingredient compared with substitutes such as domestic Chinese corn, soybean meal, distillers grains produced in China, and imported corn. From June through December 2011, the average price of imported DDGS in China was 1,985 yuan/mt (about $310), which was 19 percent lower than the cost of domestic Chinese corn and 35 percent lower than soybean meal (figure 4).

Table 2a. Estimated current inclusion rates of DDGS in feed rations

	China	United States
	Percent	
Dairy	20-30	10-20
Beef	NA	10-40
Swine	10-12	10-50
Poultry	5-10	5-10

NA = No data available.

Source: U.S. estimates from Hoffman and Baker (2011) and correspondence with Shurson. China estimates from *Nongcai Baodian* (2011).

Table 2b. Total DDGS consumption by species, crop year 2010/11

	China	United States
	Percent	
Dairy	10	30
Beef	NA	55
Swine	20	10
Poultry	60	5
Aquaculture	10	NA
Total	100	100

NA = No data available.

Source: U.S. estimates from Hoffman and Baker (2011). China estimates based on reports from industry sources.

The price of U. S. DDGS has historically been related to the price of U. S. corn, largely because corn is the chief raw material and also because corn and DDGS both serve as primary energy sources in animal feeds. An increase in the price of U. S. corn usually leads to an increase in the DDGS price.

However, the rapid growth in ethanol production from late 2008 through late 2010 expanded the supply of U. S. DDGS, contributing to weaker DDGS prices relative to corn by the end of this period (figure 5). By 2011, the DDGS price was typically about 80-90 percent of the corn price, with seasonal fluctuations in the United States characterized by lower prices during the summer months when greater cattle pasture is available, followed by higher prices over the winter months when pasture is limited. For the Chinese feed industry, the decline in U. S. DDGS prices relative to corn made DDGS a more attractive alternative to importing U. S. corn.

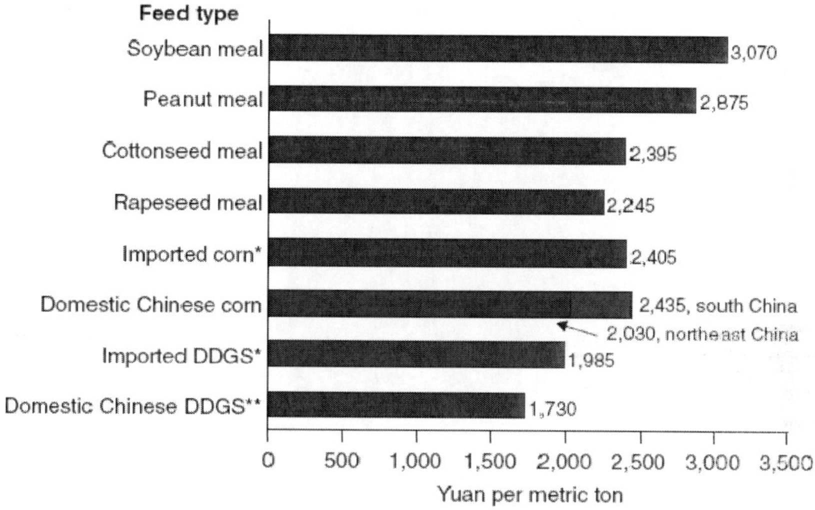

DDGS = Distillers dried grains with solubles.

*Calculated from average value of commodities reported in China customs statistics plus tariff (1 percent on corn, 5 percent on DDGS) and value-added tax (VAT) (13 percent on corn, 0 percent on DDGS).

**Price in northeast China.

Source: China National Grain and Oils Information Center, unpublished reports, except where noted.

Figure 4. Average prices of feed ingredients in China, June-December.

In addition to the attractive price of U. S. DDGS, importers in Asia are attracted by the golden yellow color of U. S. DDGS and their consistency, which reflects the uniformity of raw materials and production processes used by the U. S. ethanol industry. [6] The color is also indicative of other nutritive content, including ash, fiber, crude fat, and lysine. Research has shown that dark-colored DDGS has a lower nutritional value, primarily the result of excessive drying (Cromwell et al., 1993; Fastinger et al., 2006). Some Chinese feed compounders report that they value the color of U. S. DDGS for its influence on the color of the final feed product (Tuan et al., 2008).

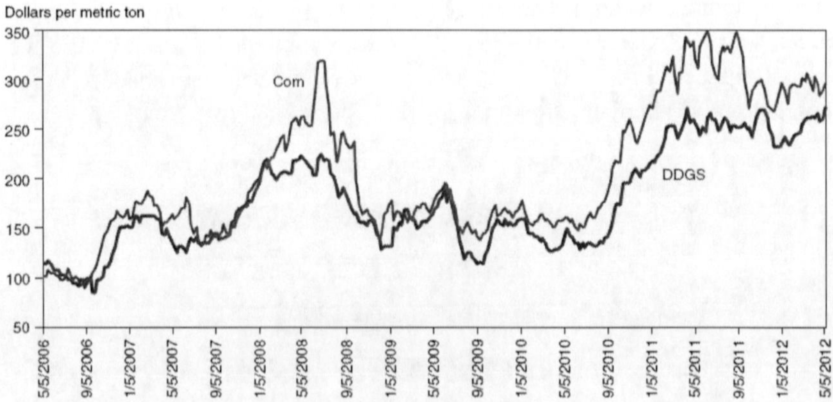

Note: Prices are on a dry-matter basis, which assumes 10-percent moisture for DDGS and 15-percent moisture for corn. DDGS = Distillers dried grains with solubles.
Source: USDA, Agricultural Marketing Service.

Figure 5. Central Illinois corn and DDGS prices.

Favorable Policy Treatment

Chinese tariff and tax policies play an important role in supporting imports of DDGS over other feed ingredients. Imports of DDGS are subject to a relatively low 5-percent tariff, but most importantly are exempt from value added taxes (VAT) and are not subject to import quotas (table 3).

Imports of corn—considered a strategic commodity in China—are regulated by a tariff rate quota system established when China became a World Trade Organization member in 2001. Corn imports within the annual quota of 7. 2 mmt are subject to a tariff of only 1 percent, but 60 percent of the quota is reserved for state trading companies.[7] Other companies must go through cumbersome procedures to apply for a share of the 2. 9 mmt available to non-state trading companies who are importers.[8] The over-quota tariff rate for corn is 65 percent ad valorem, and is effectively prohibitive. DDGS imports are not subject to an import quota, which simplifies import procedures and gives importers more flexibility in making import decisions.

DDGS imports also receive favorable tax treatment. Imported corn and soybean meal are assessed a 13-percent VAT on the landed cost of the commodities.[9] Chinese regulations exempt most imported feed ingredients like DDGS from the VAT, but the exemption for soybean meal was removed in 1999 (see box, "Favorable Treatment of Imported Feeds").

Table 3. China tariff and value added tax on selected feed ingredients

Item	Value added tax (VAT)	Tariff	Total	Import quota
		Percent		
Corn	13	1	14	Yes
Soybeans	13	3	16	No
Soybean meal	13	5	18	No
DDGS	0	5	5	No

DDGS = Distillers dried grains with solubles.
Source: China Ministry of Finance and China Customs.

Transport and Logistics

The price of DDGS arriving in China is also influenced by ocean and rail freight costs. According to industry sources (Guangzhou Renyi, 2010), transportation from the United States to China accounts for approximately 40 percent of the cost of imported DDGS. These rates can vary from year to year. For example, during the 2008 financial crisis, freight rates to China plunged.

Most DDGS shipments are priced on a flat free-on-board basis, which gives the importer responsibility for freight costs. The trade in U. S. DDGS benefits from reduced shipping costs by utilizing the plentiful supply of empty shipping containers returning to China from the United States. While most grain exports from the United States are shipped in bulk, about 70 percent of U. S. DDGS exports to China are shipped in containers, mostly from ports on the West Coast or Gulf Coast.

Infrastructure improvements at U. S. ports could facilitate additional DDGS shipments to China. A new export facility in Longview, WA, operated by EGT Development and opened in February 2012, is expected to reduce the cost of bulk shipments to Asian export markets such as China. However, the facility's effect on DDGS trade could be limited since only a few Chinese traders import in bulk. Many Chinese importers prefer containerized shipments that facilitate easier separation and handling of relatively small shipments. DDGS can at times be difficult to handle and unload because the high fat content makes DDGS clump and stick to the sides of containers (Ferris, 2011).

Favorable Treatment of Imported Feeds

Since the 1990s, Chinese authorities have experimented with trade policies designed to augment China's feed supply as a means of facilitating growth in the country's livestock industry. During a period of high feed prices in the 1990s, authorities cut tariffs on feeds derived from processing of grains, oilseeds, sugar, and alcohol distilling to 5 percent, waived the 13-percent value added tax (VAT) on imports of feeds, and eliminated import quotas.

The main purpose of these measures was to increase the supply of soybean meal to promote efficient livestock production (Crook et al., 1998). At the time, imports of unprocessed soybeans were subject to quotas, a 3-percent tariff (for shipments within the quota), and a 13-percent VAT. Soybean oil imports were also subject to a quota, a 13-percent tariff, and a 17-percent VAT.

Spurred by the waiver of VAT and removal of quotas, plus a decline in world prices, China's imports of soybean meal surged from negligible amounts in 1995 to 3. 6 mmt in 1996/97 and 4. 2 mmt during 1997/98 (Hsu, 2001). The surge of imports put downward pressure on domestic Chinese soybean meal prices and reduced profit margins of domestic soybean-crushing enterprises. At the same time, large volumes of soybean oil were smuggled into China to evade quotas, tariffs, and the VAT (Crook et al., 1998; Hsu, 2001). In July 1999, the VAT on imported soybean meal was restored in order to slow imports and boost margins for domestic soybean crushers. Since 2001, regulations have specifically excluded soybean meal from the VAT exemption for feeds. Other feeds, including DDGS, still have the VAT exemption and a 5-percent tariff.

After the VAT on soybean meal imports was restored, China shifted from importing soybean meal and soybean oil to importing mostly unprocessed soybeans. Robust investment in soybean processing and removal of quotas on imported soybeans further encouraged imports. The volume of imported soybeans soared to 54 mmt in 2011. Nearly all of the soybean meal derived from imported soybeans is used for animal feed in China. Soybean meal imports dropped to 100,000 metric tons (mt) during 2000/01 and rarely exceeded 200,000 mt annually in subsequent years.

Similar to the 1990s soybean meal episode, the surge of DDGS imports to 3 mmt during 2010 prompted complaints from domestic Chinese processors (of ethanol and alcohol) that imports depressed the price of their co-products and reduced their profit margins.

A group of these companies filed a petition with China's Ministry of Commerce on November 16, 2010, to launch an antidumping investigation against imports of U. S. DDGS (discussed in *Transports and Logistics* section of this chapter).

CHINA'S DDGS IMPORTS ARE SENSITIVE TO PRICES

Industry sources in China report that cost savings are the main motivation behind increased use of imported DDGS in China. A livestock specialist with the U. S. Grains Council's Beijing Office estimated that including DDGS could reduce feed costs for Chinese farms by 45 to 75 yuan ($7 to $11) per metric ton compared with a corn-soybean meal formulation. The manager of a Chinese feed mill asserted that cost savings were the reason for the appeal of DDGS (*Nongcai Baodian*, 2011).[10]

A study by Fabiosa et al. (2009) used actual and secondary data to conduct a microeconomic analysis on the impact of imported DDGS on feed costs in China. Their results indicated a clear economic incentive for feed millers and livestock producers to use imported DDGS in their feed rations, with a potential savings of $1 per hundredweight for mixed feed, or 6 percent of average feed costs. At the time of the study, China's DDGS imports from the U. S. were negligible, but the authors noted the potential for growth was substantial given the economic incentives alone.

The substitutability among DDGS, corn, and other feed raw materials from both domestic Chinese and imported sources means that China's imports of DDGS and corn are sensitive to relative prices and are variable, depending on the ration-formulation methodology used. The rising price of domestic corn in China is prompting feed compounders and livestock producers to seek out substitutes for corn like U. S. DDGS. In September 2011, China's corn price was approximately 20 percent higher than a year earlier. This rate of increase was approximately three times faster than the growth in China's consumer price index (CPI). The prices of other grains and feed resources are also rising, and the resulting cost pressures on livestock producers and feed mills induce them to search for alternative feeds that can minimize their production costs.

The relationship between prices and imports of DDGS and corn is illustrated in figure 6, showing China's monthly imports of DDGS and corn (both predominantly from the United States) from 2003 to 2012. The chart also displays three prices:

- The average price of domestic Chinese corn in Guangzhou, a corn-deficit region of southern China that typically has among the highest corn prices in China.
- The estimated landed cost of importing U.S. corn, calculated by adding monthly ocean-freight rates to the average price of U. S. corn at Gulf of Mexico ports plus Chinese tariffs and taxes.
- The average cost and freight (C and F) price of imported DDGS arriving in China, calculated by dividing the monthly value of DDGS imports reported in Chinese customs statistics by the volume of imports.[11]

China's imports of both DDGS and corn were insignificant during 2007-2008 when U. S. corn prices and freight rates were at high levels. During this period, China's corn prices were lower than world prices. During mid-2008, world corn prices and freight rates plunged dramatically. However, China's demand for corn also fell during this period. The Chinese Government launched a macroeconomic stimulus program to revive industrial output in early 2009 and authorities also purchased domestic corn to support prices. DDGS and corn imports remained insignificant through the first half of 2009 even though the cost of imported corn fell to a level below the cost of domestic corn.

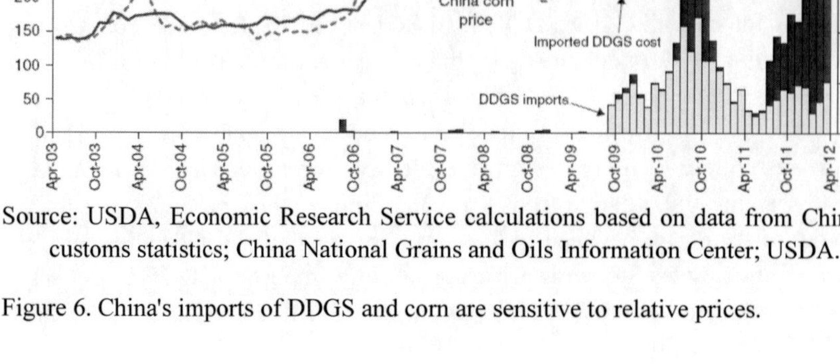

Source: USDA, Economic Research Service calculations based on data from China's customs statistics; China National Grains and Oils Information Center; USDA.

Figure 6. China's imports of DDGS and corn are sensitive to relative prices.

DDGS imports grew to significant volumes in late 2009 as the Chinese economy recovered from a brief slowdown at the end of 2008. By September 2009, imported corn was still comparable in cost to domestic Chinese corn, but the cost of imported DDGS (including the 5-percent tariff) was about 20 percent less than the average price of Chinese corn and 4 percent less than the estimated cost of imported corn (including the 1-percent tariff).[12]

A growing stream of monthly DDGS imports (nearly all from the United States) began to arrive in China in September 2009 and continued into 2010. China's corn imports remained minimal until mid-2010, when the cost of importing corn dipped and was comparable to the cost of imported DDGS. This prompted a surge of corn imports during June-September. However, that surge was choked off by an increase in the cost of imported corn during the second half of 2010.

During China's October 2009-September 2010 corn market year (2009/10), China's imports of DDGS totaled 3. 1 mmt, and imports of corn totaled 1. 3 mmt.

China's DDGS imports remained strong through December 2010, but corn imports dropped off as U. S. prices rose and the new Chinese corn harvest arrived on the market. The cost of imported DDGS remained about 15 to 25 percent below the price of Chinese corn, but the cost of imported corn rose above the domestic corn price. Consequently, China's imports of corn were again minimal during late 2010. Corn imports resumed during the second half of 2011 as U. S. prices declined and Chinese prices kept rising.

The cost of imported DDGS remained well below the cost of Chinese corn during the first half of calendar year 2011, but an antidumping investigation against U. S. DDGS announced by China's Ministry of Commerce in December 2010 raised uncertainties for Chinese purchasers (see box, "China's Antidumping Investigation Against U. S. DDGS"). DDGS imports resumed their growth during the second half of 2011 as the cost dipped and rumors circulated in China that the Government would discontinue the antidumping investigation.

During the 2010/11 market year, China's imports of DDGS fell to 2. 1 mmt and corn imports fell to 980,000 mt as the cost of imported corn rose above domestic prices while imports of rapeseed meal from Canada and India increased due to competitive prices. [13] Imports of both corn and DDGS were strong during late 2011 and early 2012 as Chinese corn prices rose and U. S. prices fell.

China's Antidumping Investigation Against U.S. DDGS

In late December 2010, China's Ministry of Commerce announced an antidumping investigation against imported U. S. DDGS to take place during 2011. The investigation was requested by a group of domestic Chinese alcohol producers that included two of China's five officially sanctioned fuel ethanol companies (Jilin Fuel Ethanol Co. and Anhui Fengyuan Biological) and two other alcohol producers in Jilin Province. There may have been other companies with an interest in the case, since China has dozens of companies producing alcohol products. The petitioners claimed that imported DDGS depressed prices of their DDGS and other co-products sold in the domestic market, thus reducing their supplementary income from alcohol production.

The potential imposition of high countervailing duties as a result of the investigation would harm Chinese feed manufacturers and livestock producers who use imported DDGS. In his role as a representative to the National People's Congress, the chairman of China's largest feed company publicly opposed the antidumping investigation, saying it was not in the interest of the feed industry and only represented the interests of a few alcohol producers ("New Hope Asks Chinese Government To Stop Antidumping Investigation Against U. S. DDGS," April 15, 2011; available at: http://www. gdfeed. org. cn/news/qiyedongtai/2011/ 0415/8985. html.) He argued that assessing high duties on imported DDGS would raise costs for feed companies that would be passed on to consumers in higher meat prices.

On June 21, 2012, the Ministry of Commerce announced the antidumping investigation would be terminated, citing a decision by the petitioners to withdraw their complaint. Some feed mill managers and analysts speculated that the investigation was launched as a strategic measure to slow the rapid growth in imported DDGS and prevent their share of the Chinese market from exceeding 50 percent (*Nongcai Baodian*, 2011).

Nonprice Factors Affecting Chinese DDGS Demand

Users may be willing to pay a premium for U. S. DDGS due to its better quality and consistency. Industry sources say DDGS produced by Chinese ethanol plants are uneven in quality and color due to the variation in raw materials and different milling and fermentation processes used (*Nongcai Baodian,* 2011). In most cases, the nutrient quality of China-produced DDGS is lower and less consistent than U. S. DDGS, reflecting differences in the ethanol production process and excessive drying, which may be the result of poor quality feedstock (USDA, Foreign Agricultural Service, 2009).

Mycotoxins in grain are a concern in China's livestock and feed industry, and the concern is even more acute for users of DDGS, particularly for poultry and swine, which are more sensitive than ruminants. High moisture levels can promote growth of molds and fungi that emit poisonous toxins that can weaken livestock and make the animals more vulnerable to disease and slow weight gain. [14] Few Chinese farmers have access to grain-drying equipment, and grain is frequently stored in facilities where it is vulnerable to moisture and temperature changes. When grain is processed into alcohol, the concentration of mycotoxins in DDGS is two to three times higher than the concentration before the starch is removed from the grain (Shurson, 2009). Industry sources say mycotoxins are a common problem in domestic Chinese DDGS, which requires costly treatment to eliminate (*Nongcai Baodian*, 2011).

On the other hand, the attraction of DDGS for Chinese buyers could be eroded by a "back end" fractionation process being adopted by U. S. ethanol producers, which raises revenue by extracting corn oil that can be used for biodiesel fuel (Ferris, 2011). Recent data from the U. S. Grains Council (2012) showed 90 out of 200 dry-mill ethanol plants in the United States currently have oil extraction capability, and the number is expected to increase to 105 by the summer of 2012. The reduced-oil DDGS produced by this process appear to have relatively greater value for cattle than for poultry and swine because ruminants can better utilize the higher fiber for energy than can monogastric animals (Shurson, 2006). [15] Higher fiber content may discourage use for Chinese poultry and swine feed—currently the dominant uses.

At a China feed industry seminar held in 2011, University of Minnesota Professor Jerry Shurson explained in a presentation that the new process reduces the fat content and a participant in the meeting questioned whether the new type of DDGS would be attractive as a feed for China's poultry and swine (Southern Rural News, 2011). So far, the effect of the reduced fat content is not clear. The relatively small share of DDGS in final feed products mutes changes in its characteristics, and some research indicates that the protein quality—the amino acid balance—of DDGS in general could keep even low-oil DDGS attractive to Chinese buyers.

The regional availability of DDGS in China is also important. Domestic DDGS and other co-products are mainly produced in the northeastern region, where corn supplies are abundant. China's southern region produces large volumes of manufactured feed products but has limited supplies of raw materials. Corn and DDGS from northeastern China must be shipped long distances at high cost to reach these customers. Guangdong Province has the largest commercial feed output in China, but corn prices there are 12 to 15 percent higher than in northeastern provinces. Consequently, southern China is the leading destination for imported DDGS. About two-thirds of imports arrived at ports in southern China. Districts in Guangdong and neighboring Guangxi Province accounted for 1. 6 mmt of DDGS imports during 2010 (figure 7). Shanghai and neighboring districts—also considered part of southern China—accounted for 0. 5 mmt. The Qingdao District accounted for 1 mmt of imports. Qingdao is considered part of northern China, but the district includes Shandong Province, the second-leading feed-producing region with output nearly matching that of Guangdong in 2010. There were no significant DDGS imports by districts in the northeast, a surplus supply region for corn where most of China's domestic DDGS originate.[16]

China's Domestic Production of Alcohol Co-products

China's antidumping investigation is a reminder that imported DDGS also compete with China's substantial domestic supply of DDGS and related residual materials from alcohol distilling. Industry sources speculated that the antidumping investigation was launched to prevent imports from gaining more than 50 percent of China's DDGS market (*Nongcai Baodian*, 2011; China Corn Net, 2011).

Fuel ethanol is part of a larger alcohol industry that yields various feed co-products. During the early 2000s, production of alcohol for beverage, fuel, pharmaceutical, and industrial uses expanded rapidly (Gale et al., 2009). The raw material used for distilling alcohol varies from region to region and includes corn, rice, wheat, sorghum, potatoes and other nongrain feed ingredients such as cassava. However, much of the increased capacity was constructed in corn-growing regions and uses corn as the primary raw material.

Corn-based co-products from domestic Chinese fuel ethanol and other types of alcohol industries are produced mainly in the northeastern region of the country, the region with the most abundant corn resources. There is no means of directly measuring the availability of residues from alcohol distilling in China, but limited statistics on alcohol production show that the industry has grown most rapidly in corn-producing regions, suggesting that the alcohol industry's growth represents increased use of corn as the raw material (Chinese Academy of Agricultural Mechanization Sciences, 2011).

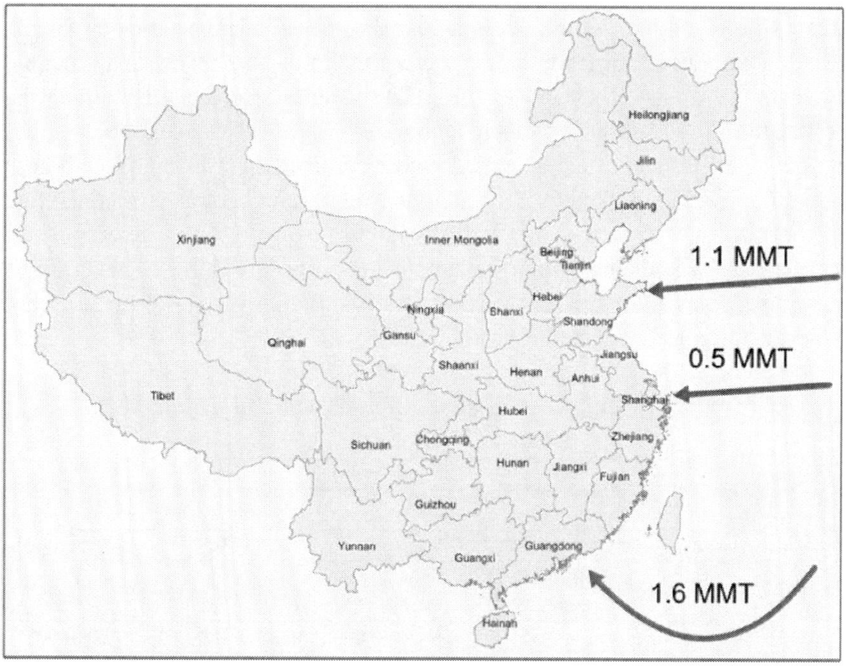

Note: Map shows volume of imports by customs district. Guangdong includes Huangpu, Guangzhou, Shenzhen, Zhanjiang, Xiamen, and Nanning customs districts. Shanghai includes Shanghai, Nanjing, and Ningbo. Shandong includes Qingdao and Tianjin.

MMT = million metric tons.

DDGS = Distillers dried grains with solubles.

Source: USDA, Economic Research Service calculations based on data from China's customs statistics; China National Grain and Oils Information Center; USDA.

Figure 7. China's imports of DDGS by region, 2011.

Co-products from fuel ethanol producers are the closest substitutes for imported DDGS. The Chinese Government has officially authorized five plants to produce fuel ethanol (see box, "Ethanol Production in China"). There are no statistics on use of raw materials by these plants. If the four grain-based ethanol plants used corn exclusively to produce at capacity (the fifth plant uses cassava), they would use roughly 5. 5 mmt of corn and have co-product residual output of about 1. 7 mmt tons (table 4).[17]

Although the reported capacity of the five authorized fuel ethanol processors is 1. 89 mmt, one article in the Chinese news media reported that fuel ethanol production could be even larger than the authorized volume (Zhang, 2010). This article quoted industry sources claiming that many small alcohol distillers reformulate their product to make unauthorized fuel ethanol and the designated producers exceed their authorized volumes when gasoline prices are high. The article's findings could not be confirmed but listings on Chinese e-commerce sites checked by ERS include a number of companies listing "fuel ethanol" as one of their products. This could also imply a larger supply of domestic DDGS than is often acknowledged. Recent Chinese Government policy has sought to slow growth of ethanol and other industrial uses of corn but production has remained robust.

Table 4. Officially designated ethanol plants in China, 2010

Company	Primary inputs	Capacity	
		Metric tons	*Gallons*
Jilin Fuel Alcohol Company Limited	Corn	550,000	183,380,000
Henan Tianguan Group	Wheat, Corn, Cassava	500,000	166,710,000
Anhui BBCA Group	Wheat and Corn	420,000	140,040,000
Heilongjiang COFCO	Corn	220,000	73,350,000
Guangxi COFCO	Cassava	200,000	66,680,000
Assuming corn is the only input in Henan and Anhui:			
Total corn ethanol production		1,690,000	
Total corn used		5,070,000	
Total DDGS		1,690,000	

DDGS = Distillers dried grains with solubles.

Source: China Petroleum and Chemical Corporation (Sinopec).

Ethanol Production in China

China began its ethanol program in 2001 when the Chinese Government designated four fuel ethanol producers in Jilin, Heilongjiang, Henan, and Anhui Provinces that used corn, rice, and wheat as raw materials.

At the time, the Government held large reserves of grain and farmers were receiving low prices for grain. The four ethanol producers receive cash subsidies and tax exemptions and sell their output to two state-owned fuel blenders.

In 2006, the Chinese Government announced that new growth in ethanol output would come from nongrain raw materials such as cassava, sweet sorghum, sweet potatoes, crop residues, and forest products. The nongrain raw materials can only be planted on land that is unsuitable for producing grain and processing must be nonpolluting. In 2009, a fifth plant began producing ethanol in Guangxi Province using cassava as a raw material. Several ethanol producers conducted experimental trials in producing ethanol from sweet sorghum and corn stalks. The feed value of co-products from nongrain raw materials is generally low. An Economic Research Service team that visited the Guangxi plant was told that the residual by product from producing ethanol from cassava can be used as animal feed, but it has a lower protein content (11-14 percent) and higher fiber (30-40 percent). The company burns most of the residual by product for electricity generation (Tuan et al., 2008).

China's 12th 5-year plan for 2011-15 announced a target of producing 5. 0 mmt of biofuels by 2015, with all growth coming from nongrain raw materials. In 2011, China's State Council approved a plan to construct a plant in eastern China's Zhejiang Province that will use cassava imported from Southeast Asia as the raw material. Other nongrain ethanol projects are planned, and research on new technologies continues.

In 2012, China's Ministry of Finance sharply reduced the subsidy for grain-based ethanol production. The subsidy for 2012 was set at 500 yuan/mt, less than half the 1276 yuan subsidy in 2011 (the 2012 subsidy for nongrain fuel ethanol was set at 750 yuan/mt). The reduction reflects concerns among Chinese officials that rising use of corn for industrial processing was contributing to upward pressure on corn prices. The shift in policy is likely to restrict the domestic supply of DDGS by slowing or reducing corn-based ethanol production in China.

CONCLUSION

China's robust demand for feed ingredients makes the country an important market for U. S. DDGS exports, but the volume of U. S. DDGS exported to China in the future is highly uncertain. A key constraint on future growth may be prospects for slower expansion in the supply of U. S. DDGS during the coming decade, as growth in the volume of corn-based ethanol produced is expected to slow. According to the latest USDA baseline projections to 2021, total U. S. corn used for ethanol is forecast to increase from 128 mmt in 2010/11 to 139 mmt in 2021/22, compared with growth of 112 mmt between 2000/01 and 2010/11 (USDA/OCE, 2012). This moderation of growth reflects slower expected increases in blended gasoline consumption in the United States and constraints in the market for ethanol blends above 10 percent (E10), including approved E15 and E85 blends (USDA/OCE, 2012). Based on this projected level of ethanol output, U. S. DDGS supplies will likely expand, but at a much more subdued rate than in recent years.

It is possible that the volume of U. S. DDGS available for export will decline if U. S. domestic use of DDGS increases in the future. DDGS has become an important component of U. S. livestock rations, and domestic feed use is unlikely to fall substantially from an estimated 29 mmt in 2010/11. 18 The projected increase in U. S. meat consumption and exports over the next decade could also result in further increases in domestic demand for DDGS (USDA/OCE, 2012). Therefore, it is possible U. S. domestic use of DDGS could expand faster than production, but this will depend, in part, on the relative availability and prices for corn and soybean meal.

U. S. DDGS supply constraints aside, relative prices and transportation costs will likely have the biggest influence on the destination of U. S. DDGS exports in the future. High transportation costs of DDGS and integration of feed and livestock sectors in the North American Free Trade Association region favor nearby markets—such as Canada and Mexico—over Asian markets. [19] On the other hand, high feed prices and robust domestic demand in other countries could push DDGS export sales toward other growth destinations in Asia besides China, such as South Korea, Vietnam, and Thailand.

Use of imported DDGS in China is a relatively new phenomenon and limited to a few relatively small feed mills in coastal provinces. The cost advantage depends in large part on trends in U. S. and China corn prices including tariffs and taxes. If China's corn prices continue rising, feed mills and livestock producers will face continued pressure to find cheaper feed ingredients like DDGS. In the United States, DDGS and corn prices tend to be closely related, so a decline in U. S. corn prices relative to Chinese corn prices would tend to make both U. S. corn and DDGS more price-competitive in China.

Chinese policy treatment of DDGS is uncertain as officials balance the priority of expanding feed supplies against domestic corn-processing industry interests. Favorable trade treatment of DDGS is an important factor that spurred the import of DDGS. However, the antidumping investigation against U. S. DDGS reflects the competing interests of livestock producers and domestic suppliers of DDGS (China Corn Net, 2011). DDGS imports affect only a relatively small number of alcohol producers in China, but industry sources indicate that policymakers and industry members are averse to imports gaining a dominant share of any industry (*Nongcai Baodian*, 2011). [20] The antidumping investigation was terminated in 2012, suggesting that concerns about feed costs prevailed over alcohol industry interests. At the same time, Chinese officials have sharply reduced subsidies for domestic fuel ethanol and implemented other policies to constrain growth in domestic alcohol production that would restrict domestic supplies of DDGS and other co-products that compete with imported DDGS.

China's feed demand is expected to expand rapidly over the next 10 years. USDA's baseline projections to 2021 project that feed and residual use of corn will expand from 124 mmt to 172 mmt and soybean meal consumption will rise from 43 mmt to 73 mmt. This expansion in feed use is driven by a livestock sector shifting from backyard to commercial production, which will likely utilize more corn and soybean meal and fewer inputs such as household and processing waste, crop by-products, vines, tubers, and other forages.

Feed demand in China will remain robust, but future DDGS sales to China depend on the economic value of U. S. supplies compared with alternative raw materials including domestic and imported corn. DDGS from the United States are one of many sources of raw materials supporting China's growing commercial feed production (imports of DDGS were equivalent to roughly 2 percent of feed output during 2010). In an environment of rising costs, Chinese feed mills will continue to seek out cost-minimizing raw materials. If DDGS sales to China are restricted by rising prices or other factors, Chinese demand will shift to other raw materials: corn, wheat, other grains, oilseed meals, and other milling and processing co-products. The emergence of China's DDGS imports is a reminder that feed supplies come from diverse sources, and demand is price-sensitive.

References

Berger, Larry L., and Darrel L. Good. 2007. "Distillers Dried Grains Plus Solubles Utilization by Livestock and Poultry," Corn-Based Ethanol in Illinois and the U.S.: A Report from the Department of Agricultural and

Consumer Economics, University of Illinois, Chapter 6, November. Retrieved from: http://www. farmdoc. illinois. edu/policy/research_ reports/ethanol_report/Ethanol%20Report. pdf

Bunge North America. 2011. EGT To Build Third Facility in Montana. August 16. Retrieved from: http://www. bungenorthamerica. com/news/ releases/2011-08-16. shtml

China National Grains and Oils Information Center (CNGOIC). Feedgrains Supply and Demand Situation Weekly Report. Various issues.

Chinese Academy of Agricultural Mechanization Sciences. 2011. "2010 China Agricultural Processing Yearbook," China Agricultural Press, July.

China Aquaculture Network. 2010. Zhongguo jiang jinkou geng duo yumi gan jiuzao danbai siliao (China will import more DDGS), November 19. Retrieved from: http://www. yangzhiwang. com. cn/news/show. php?itemid=3257

China Corn Net. 2011. DDGS fan qing xiao diaocha yinfa si kao (Thoughts on what triggered the DDGS antidumping investigation), news report, January 10. Retrieved at: www. yumi. com. cn

China Feed Trade. 2011. Zhong mei siliao duiyu yumi de xuqiu jiang hui chixu shangsheng (China and U. S. corn feed demand will continue to increase), September 23. Retrieved from: http://www. xumu. com. cn/html/ feed/1011/2011/0926/232407. html

China Ministry of Agriculture. 2011. "2010 China Agricultural Yearbook," China Agricultural Press, March.

China Ministry of Agriculture. 2011. China's 12th 5-year Plan for the Feed Industry, retrieved from: http://www. moa. gov. cn/zwllm/zcfg/ nybgz/201110/t20111019_2367679. htm

China Ministry of Agriculture. 2011. China's 12th 5-year Plan for the Fisheries Industry, retrieved from: http://www. moa. gov. cn/zwllm/ ghjh/201111/t20111114_2408578. htm

China Ministry of Agriculture. 2011. China's 12th 5-year Plan for the Live-stock Industry, retrieved from: http://web. zgny. com. cn/User_Doc/ 23530/ News/2422/20119301138115663235301. htm

China Ministry of Agriculture. 2006. China's 11th 5-year Plan for the Live-stock Industry, retrieved from: http://www. gov. cn/gzdt/2006-09/13/ content_387510. htm

China Ministry of Agriculture Feed Industry Office. 2011. 2010 Quanguo Nian Siliao Gongye Shengchan (2010 National Feed Industry Production), news release, May 19.

Coffing, Art, Hunter Colby, William Lin, and James R. Simpson. 1991. China Trip Report (U.S. Livestock and Feed Team), USDA Science and Technology Exchange Program internal report, October.

Cromwell, Gary, Kevin Herkelman, and Tim Stahly. 1993. "Physical, Chemical and Nutritional Characteristics of Distillers Dried Grains With Solubles for Chick and Pigs," Journal of Animal Science 71(3):679- 686, March 1. Retrieved from: http://jas. fass. org/content/71/3/679. full. pdf+html

Crook, F. W., R. Hanson, H. Hsu, B. Karmen, and P. Laney. 1998. Soybean Consumption and Trade in China: A Trip Report, USDA internal report, July.

Dooley, Frank J. 2008. U.S. Market Potential for Dried Distillers Grains With Solubles, Working Paper #08-12, Purdue University, Department of Agricultural Economics, December. Retrieved from: http://ageconsearch. umn. edu/bitstream/46342/2/08-12. pdf

Fabiosa, Jacinto F., James Hansen, Holger Matthey, Suwen Pan, and Francis Tuan. 2009. Assessing China's Potential Import Demand for Distillers Dried Grains: Implications for Grain Trade, Staff Report 09-SR 104, Iowa State University, Center for Agricultural and Rural Development, December. Retrieved from: http://www. card. iastate. edu/publications/DBS/ PDFFiles/09sr104. pdf

Fastinger, N. D., J. D. Latshaw, and D. C. Mahan. 2006. "Amino Acid Availability and True Metabolizable Energy Content of Corn Distillers Dried Grains With Solubles in Adult Cecectomized Roosters," Poultry Science 85:1212-1216, 2006. Retrieved from: http://ps. fass. org/cgi/reprint/85/7/12 12?maxtoshow= and hits=10 and RESULT FORMAT= and author1=Fastinger and an dorexacttitle=and and andor exacttitleabs=and and andorexactfulltext=and and sear chid=1 and FIRSTINDEX=0 and sortspec=relevance and fdate=1/1/2003 and tdate=9 /30/2008 and resourcetype=HWCIT

Ferris, John. 2011. Potential for Corn Oil Extracted from Distillers Dried Grains and Solubles as a Feedstock for Biodiesel, Staff Paper 2011-07, Michigan State University, September.

Gale, F., Marti, D., and Hu, D. 2012. China's Volatile Pork Industry, LDP-M-211-01, USDA, Economic Research Service, February. Retrieved from: http://www. ers. usda. gov/media/262067/ldpm21101_1_. pdf

Gale, F., and S. Henneberry. 2010. "Markets Adapt to China's Changing Diet," Choices 24, 2nd Quarter 2010, pp. 32-36.

Gale, F., Tuan, F., Wang, X., and Cao, Z. 2009. China Is Using More Corn for Industrial Products, FDS-09k-01, USDA, Economic Research Service, December. Retrieved from: http://www. ers. usda. gov/media/153522/ fds09k01. pdf

Guangzhou Renyi Import and Export Trading Company Limited. 2010. Jinkou meiguo DDGS jingyan fenxiang (Our experience importing U. S. DDGS), presentation given at U. S. DDGS promotion conference in China. Retrieved from: http://122. 70. 144. 155/ppt/

Hoffman, Linwood, and Allen Baker. 2011. Estimating the Substitution of Distillers Grains for Corn and Soybean Meal in the U.S. Feed Complex, FDS-I-01, USDA, Economic Research Service, October. Retrieved from: http://www. ers. usda. gov/media/236568/fds11i01_2_. pdf

Hoffman, Linwood, and Allen Baker. 2010. Market Issues and Prospects for U.S. Distillers Grains: Supply, Use, and Price Relationships, FDS-10k01, USDA, Economic Research Service, December. Retrieved from: http://www. ers. usda. gov/media/107533/fds10k01_1_. pdf

Hsu, Hsin-Hui. 2001. "Policy Changes Continue to Affect China's Oilseeds Trade Mix," in China: Agriculture in Transition, WRS-01-2, USDA, Economic Research Service, November, pp. 30-36.

Jacela, J. Y., Frobose, H. L., DeRouchey, J. M., Tokach, M. D., Dritz, S. S., Goodband, R. D., and Nelssen, J. L. 2010. "Amino acid digestability and energy concentration of high-protein corn dried distillers grains and high-protein sorghum dried distillers grains with solubles for swine," Journal of Animal Science 89:3617-3623, 2010. Retrieved from: http://jas. fass. org/content/88/11/3617. full. pdf+html

Lao, Jinhua. 2009. "Siliao Meijian Dusu Wuran Yanzhong, Shangjia de Xuetou? (Is serious mycotoxin contamination in feed a commercial stunt?)," Southern Rural News, July 17. Retrieved from: http://nf. nfdaily. cn/nanfangdaily/nfncb/200907160241. asp

National Corn Growers Association. 2008. Value-added Feed Source for Beef, Dairy Beef, Dairy, Poultry, Swine, and Sheep, September 9. Retrieved from: http://www. ethanol. org/pdf/contentmgmt/NCGA_ Distillers_ Grain_908-1. pdf

Nongcai Baodian. 2011. "Wu Nian Nei DDGS Jinkou Chao Guochan, Shangwu Bu Li An Fan Qingxiao Diaocha (DDGS Imports Surpassed Domestic Production in Five Years, Ministry of Commerce Launched Antidumping Investigation)," news article, April 7. Retrieved from: http:// animal. nfncb. cn/sczlzt/32. html

Shurson, Jerry. 2002. The Value and Use of Distiller's Dried Grains with Solubles (DDGS) in Swine Diets, presentation retrieved from: http://www. ddgs. umn. edu/articles-swine/2002-Shurson-%20Carolina%20nutr%20 conf. pdf

Shurson, Jerry. 2006. "The Value of High-Protein Distillers Co-products in Swine Feeds. " Distillers Grains Quarterly, First Quarter 2006, pp. 22-25. Retrieved from: http://www. ddgs. umn. edu/articles-swine/2006-Shurson-%20DGQuarterly. pdf

Shurson, Jerry, and Abdorrahman S. Alghandi. 2008. "Quality and New Technologies to Create Corn Co-Products from Ethanol Production," chapter 10 in Using Distillers Grains in the U.S. and International Livestock and Poultry Industries, Iowa State University, Midwest Agribusiness Trade Research and Information Center at the Center for Agricultural and Rural Development. Retrieved from: http://www. card. iastate. edu/books/distillers_grains/pdfs/distillers_grains_book. pdf

Shurson, Jerry, and Mindy Spiehs. 2002. Feeding Recommendations and Example Diets Containing Minnesota-South Dakota Produced DDGS for Swine, retrieved from: http://www. ddgs. umn. edu/feeding-swine/examples wine diets-revised. pdf

Sinopec News (Zhongguo Shihua Xinwen Wang). 2011. Ranliao Yichun Chanye Fazhan Jinru Pingjing Qi (Ethanol Industry Development Reaches Bottleneck), news article, June 29.

Southern Rural News (Nanfang Nongcun Bao). 2011. Youzhi Hanliang Jianshao Huojiang DDGS Xing Jia Bi (Reduced Fat Content Could Lower the Value of DDGS), news article, November16. Retrieved at: http://www. nfncb. cn/nccf/xm/4967. html

Southern Rural News (Nanfang Nongcun Bao). 2009. Siliao Meijian Dusu Wuran Yanzhong (Serious Mycotoxin Contamination of Feed), news article, July 17.

Tuan, F., L. Hoffman, W. Lin, J. Hansen, and F. Gale. 2008. China's Potential Demand for Ethanol Co-products. USDA, Economic Research Service internal report, July.

U. S. Department of Agriculture, Agricultural Marketing Service. 2012. Livestock and Grain Market News Portal. Retrieved from: http://marketnews. usda. gov/portal/lg

U. S. Department of Agriculture, Economic Research Service. 2011. Feed Grains Database, based on August 2011 World Agricultural Supply and Demand Estimates. Retrieved from: http://www. ers. usda. gov/data/feedgrains/

U. S. Department of Agriculture, Foreign Agricultural Service. 2011. Production, Supply and Distribution Online, based on August 2011 World Agricultural Supply and Demand Estimates. Retrieved from: http://www. fas. usda. gov/psdonline/

U. S. Department of Agriculture, Foreign Agricultural Service. 2012. Global Agricultural Trade System Online. Retrieved from: http://www. fas. usda. gov/gats/default. aspx

U. S. Department of Agriculture, Foreign Agricultural Service. 2009. U.S. DDGS Dashes for China, Mostly to the Feed Mills of Guangdong, GAIN Report Number: CH9611, Global Agriculture Information Network, July.

U. S. Department of Agriculture, Office of the Chief Economist, World Agricultural Outlook Board. 2012. USDA Agricultural Projections to 2021, OCE-2011-1, February. Retrieved from: http://www. ers. usda. gov/publications/oce-usda-agricultural-projections/oce121. aspx

U. S. Grains Council. 2012. "Low-oil DDGS Become Increasingly Available," Ethanol Producer, Vol 18, Issue 2, January 26.

Wang, Enhui. 2011. "Zhongguo Siliao Shengchan Zhuangkuang ji Fazhan Qushi (China's Feed Industry Production Situation and Development Trend)," Agricultural Outlook, No. 6, 2011, pp. 43-46.

Zhang, Qian. 2010. Qianwan Dun Che Yong Yichun Chanliang Cunyi (Millions of tons of ethanol production in doubt), Caijing, August 11. Retrieved from: http://www. caijing. com. cn/2010-08-11/110495605. html

End Notes

[1] One metric ton (mt) of corn produces 106 gallons of ethanol and 0. 3125 mt of DDGS. Assuming cash prices of $2. 80/gallon for ethanol and $230/mt for DDGS, the resulting total revenue per mt of corn is (106 X $2. 8) + ($230 X 0. 3125) = $368. Therefore, $72/$368 = 20 percent.

[2] On a September-August basis, the marketing year for corn in the United States. Data source is U. S. Department of Commerce, U. S. Census Bureau. Total trade differences occur between U. S. Census and China's Customs Statistics.

[3] China's aquatic products are raised in both fresh and salt water, often using commercial feed . The 5-year plan projects the share of farm-raised aquatic products to rise from 71 to 75 percent during 2010-15 .

[4] A USanalysis also observed that USDDGS are commonly used as a substitute for corn (Ferris, 2011)

[5] These totals do not include raw grains and other feeds mixed on farms .

[6] This color preference for feed ingredients is not new. During the late 1990s, when China previously imported large quantities of soybean meal, ERS interviews of Chinese end users

indicated they disliked the dark color of Brazilian soybean meal (Crook et al., 1998). As a result, it was blended with U. S. soybean meal to improve the overall color.

[7] These are companies that import on behalf of the Government to replenish commodity reserves.

[8] Applicants for quota must meet requirements for production capacity and have a history of importing corn to request a quota allocation from Chinese Government authorities each year. China had 10,804 feed-processing enterprises in 2010 (China Ministry of Agriculture Feed Industry Office, 2011).

[9] Greater detail about the definition of "landed cost" can be found in the next chapter, "China's DDGS Imports Are Sensitive to Prices. "

[10] The Beijing Office of the U. S. Grains Council posts on its website periodic reports showing the costs and feed value of DDGS and other feed raw materials in China (see: http://www. grains. org. cn/_d269507396. htm).

[11] Note that customs statistics could not be used to calculate the C and F price of imported corn since China's corn imports were negligible in most months.

[12] The landed corn price shown in figure 6 does not include the VAT. A number of large corn sales to China appear to have been made to state trading entities, presumably to replenish Government reserves. Industry sources indicate these imports are exempt from VAT, but ERS has not been able to confirm this. Other purchases of imported corn are subject to VAT.

[13] China's imports of rapeseed meal totaled 1 .4 mmt during October 2010-September 2011, up from 1 mmt during the year-earlier period .

[14] According to one study, tests of feed products in China during 2008 found that nearly all of 342 samples were contaminated with mycotoxins, and products of corn (DDG, corn gluten meal, brewers' grains) had the highest prevalence (Lao, 2009).

[15] Anticipating this problem for a wide spectrum of users who use DDGS as a substitute for corn, Ferris (2011, p. 9) surmised that "deoiled" DDGS may sell at a discounted price for some time after adoption of "back end" fractionation.

[16] China actually exports a small volume of DDGS to South Korea and Japan, which are geographically close to northeast China. However, despite the short distance and low cost of Chinese DDGS, South Korea and Japan import DDGS predominantly from the United States. The pattern of these bilateral trade flows may be attributable in part to the low and uneven quality of Chinese DDGS.

[17] The calculation assumes 3 mt of corn yield 1 mt of ethanol and 1 mt of DDGS. There are 3. 785 liters in 1 gallon and 1,262 liters per mt of ethanol.

[18] DDGS appears to have substituted for direct use of corn and soybean meal by livestock producers following the expansion of ethanol use of corn. See Hoffman and Baker (2011).

[19] Some have claimed that as transport cost becomes a major consideration, it will favor importing feed nutrients in the form of DDGS rather than corn because the density is magnified by three times in the former. Supply constraints aside, the substitution of DDGS for corn depends on many factors and further research is needed to answer this question. Handling advantages for corn could partly offset gains from nutrient concentration in DDGS.

[20] The soybean industry is dominated by imports, and commentators warn that rising imports of DDGS and other commodities could lead to a similar reliance on imports.

INDEX

C

D

E

F

G

H

I

J

K

L

M

N

O

P

Q

R

S

T

U

V

W

Y